S0-BSK-789

Global Catastrophes: A Very Short Introduction

VERY SHORT INTRODUCTIONS are for anyone wanting a stimulating and accessible way in to a new subject. They are written by experts, and have been translated into more than 40 different languages.

The Series began in 1995, and now covers a wide variety of topics in every discipline. The VSI library now contains over 350 volumes—a Very Short Introduction to everything from Psychology and Philosophy of Science to American History and Relativity—and continues to grow in every subject area.

Very Short Introductions available now:

ACCOUNTING Christopher Nobes
ADVERTISING Winston Fletcher
AFRICAN AMERICAN RELIGION
    Eddie S. Glaude Jr.
AFRICAN HISTORY John Parker and
    Richard Rathbone
AFRICAN RELIGIONS Jacob K. Olupona
AGNOSTICISM Robin Le Poidevin
ALEXANDER THE GREAT
    Hugh Bowden
AMERICAN HISTORY Paul S. Boyer
AMERICAN IMMIGRATION
    David A. Gerber
AMERICAN LEGAL HISTORY
    G. Edward White
AMERICAN POLITICAL PARTIES
    AND ELECTIONS L. Sandy Maisel
AMERICAN POLITICS Richard M. Valelly
THE AMERICAN PRESIDENCY
    Charles O. Jones
AMERICAN SLAVERY
    Heather Andrea Williams
ANAESTHESIA Aidan O'Donnell
ANARCHISM Colin Ward
ANCIENT EGYPT Ian Shaw
ANCIENT GREECE Paul Cartledge
THE ANCIENT NEAR EAST
    Amanda H. Podany
ANCIENT PHILOSOPHY Julia Annas
ANCIENT WARFARE Harry Sidebottom
ANGELS David Albert Jones
ANGLICANISM Mark Chapman
THE ANGLO-SAXON AGE John Blair
THE ANIMAL KINGDOM
    Peter Holland

ANIMAL RIGHTS David DeGrazia
THE ANTARCTIC Klaus Dodds
ANTISEMITISM Steven Beller
ANXIETY Daniel Freeman and
    Jason Freeman
THE APOCRYPHAL GOSPELS
    Paul Foster
ARCHAEOLOGY Paul Bahn
ARCHITECTURE Andrew Ballantyne
ARISTOCRACY William Doyle
ARISTOTLE Jonathan Barnes
ART HISTORY Dana Arnold
ART THEORY Cynthia Freeland
ASTROBIOLOGY David C. Catling
ATHEISM Julian Baggini
AUGUSTINE Henry Chadwick
AUSTRALIA Kenneth Morgan
AUTISM Uta Frith
THE AVANT GARDE David Cottington
THE AZTECS David Carrasco
BACTERIA Sebastian G. B. Amyes
BARTHES Jonathan Culler
THE BEATS David Sterritt
BEAUTY Roger Scruton
BESTSELLERS John Sutherland
THE BIBLE John Riches
BIBLICAL ARCHAEOLOGY
    Eric H. Cline
BIOGRAPHY Hermione Lee
THE BLUES Elijah Wald
THE BOOK OF MORMON
    Terryl Givens
BORDERS Alexander C. Diener and
    Joshua Hagen
THE BRAIN Michael O'Shea

For more information visit our website

www.oup.com/vsi/

Bill McGuire

GLOBAL CATASTROPHES

A Very Short Introduction
SECOND EDITION

OXFORD
UNIVERSITY PRESS

AURORA PUBLIC LIBRARY

OXFORD
UNIVERSITY PRESS

Great Clarendon Street, Oxford, OX2 6DP,
United Kingdom

Oxford University Press is a department of the University of Oxford.
It furthers the University's objective of excellence in research, scholarship,
and education by publishing worldwide. Oxford is a registered trade mark of
Oxford University Press in the UK and in certain other countries

© Bill McGuire 2002, 2005, 2014

The moral rights of the author have been asserted

First published in hardback as A Guide to the End of the World 2002
First published as a Very Short Introduction 2005
Second edition published 2014

Impression: 1

All rights reserved. No part of this publication may be reproduced, stored in
a retrieval system, or transmitted, in any form or by any means, without the
prior permission in writing of Oxford University Press, or as expressly permitted
by law, by licence or under terms agreed with the appropriate reprographics
rights organization. Enquiries concerning reproduction outside the scope of the
above should be sent to the Rights Department, Oxford University Press, at the
address above

You must not circulate this work in any other form
and you must impose this same condition on any acquirer

Published in the United States of America by Oxford University Press
198 Madison Avenue, New York, NY 10016, United States of America

British Library Cataloguing in Publication Data
Data available

Library of Congress Control Number: 2014936753

ISBN 978-0-19-871593-1

Printed in Great Britain by
Ashford Colour Press Ltd, Gosport, Hampshire

AURORA PUBLIC LIBRARY

For my sons; Fraser and Jake. May their lives be long, happy, peaceful, and catastrophe-free

Contents

Preface: where will it all end?

Que será, será
Whatever will be will be
The future's not ours to see
Que será, será

'Que Será Será': Jay Livingston and Ray Evans

In a normal year on the disaster front—if there can ever be such a thing—a staggering 106,000 people, on average, lose their lives to Nature's random culls. Over the last seven years, natural catastrophes took more than 600,000 lives and cost an astonishing one trillion dollars—the 2011 Tohuku earthquake and accompanying tsunami alone ringing up a staggering US$210 billion bill—and there is little sign of any respite. On the contrary, the ever-upward trend of extreme weather events, driven by climate change, points to a future of more severe droughts, even bigger floods, and more powerful windstorms.

Notwithstanding the growing devastation and the enormous death tolls, however, society has not collapsed, nor has our planet been obliterated. No volcanic super-eruption has plunged our world into bitter cold, and comets and asteroids—barring the mini version that exploded above Siberia in 2013—have kept their distance. The prospect of another ice age has receded, while global warming has not (yet) begun to do its worst. So thankfully, despite the continuing geophysical mayhem, I am still here to write this updated version of *Global Catastrophes*, and you are still here to read it. But how long can this situation persist? Our world and our society remain fragile entities that continue to prevail solely by Nature's consent. Although it may not seem like it to those caught up in flood or drought, volcanic blast or quake, modern human civilization has developed against a background of relative geological and climatic calm. Looking back into deep time, however, it quickly becomes apparent that such stability is far from normal, so that our society owes its explosive growth and break-neck technological development to a dozen or so millennia of tranquillity shoe-horned between the last Ice Age and the one—global warming permitting—to come. In the absence of the huge quantities of greenhouse gases with which we continue to contaminate our precious atmosphere, the ice sheets would be advancing again in a few thousand years, and we would be faced with the prospect of a return to a frigid planet. As it is, due to our polluting activities, the Earth is now warming at an unprecedented rate, bringing the prospect, not of a world of ice, but of a tropical hothouse.

Nothing lasts forever—not even a planet—so the world has to end sometime. In about five billion years' time when our Sun finally runs out of fuel and swells to become a bloated red giant, it will burn the Earth to a cinder. There are also many alternative and imaginative ways in which our world or our race might meet their ends far sooner, of which disease, atomic warfare, natural catastrophe, nanobot hell, and exotic physics experiments gone wrong, are but a small selection. Given the current state of the planet you too might be forgiven for having second thoughts about the proximity of our ending. Perhaps, after all, we will face 'doom soon' as John Leslie succinctly puts it in his book *The End of the World*, rather than 'doom deferred'. Against a background of accelerating climate change, exploding population, and rampant and widespread civil strife, it may indeed be more logical for us to speculate that the human race's great adventure is about to end, rather than persist far into the future and across the vastness of galactic space.

Somewhat worryingly, Cambridge cosmologist Brandon Carter has developed an argument that supports, probabilistically, this very thesis. His 'doomsday argument' goes like this. Assuming that our race grows and persists for millions or even billions of years, then those of us alive today must belong to the infinitesimally small fraction of humans living in the earliest light of our race's dawn. This, Carter postulates, is statistically unlikely in the extreme. It is much more probable that we are alive at the same time as, say, 10 per cent of the human race. This is another way of saying that humans will cease to exist long before they have any chance to spread across space in any numbers worth talking about. John Leslie illustrates this argument along these lines. Imagine your name is in a lottery draw, but you don't know how many other names there are. You have reason to believe, however, that there is a 50 per cent chance that the total number is a thousand and an equal probability that the total is ten. When the tickets are drawn, yours is one of the first three. Now, there can be few people who, in such circumstances, would believe that the draw contained a thousand rather than ten tickets.

If the doomsday argument is valid—and it has withstood some pretty fierce attacks from a number of intellectual heavyweights— then we may have only a few centuries' respite before climate change or some other Nemesis annihilates our race, our planet, or both. Notwithstanding this, wiping out seven billion or more people at a stroke is never going to be easy, however bad things get as a consequence of mankind's destructive activities.

Furthermore, many of the imminent, so-called 'end of the world' scenarios promulgated by people who put even my doom-mongering to shame are, in reality, no such thing, but would simply result—at worst—in a severe fall in human numbers and/or the contraction of our global, technological civilization to something far simpler and more parochial—at least for a time. Personally, therefore, I am open-minded about what Stephen Baxter calls in his novel *Manifold Time* the 'Carter Catastrophe'. There is no question that the human

xvii

race or its descendants must eventually succumb to oblivion, but however depressing prospects for its short- to medium-term future seem, that time may yet be some way off.

This might be a good point to look more carefully at just what we understand by 'the end of the world' and how I will be treating the concept in this book. To my thinking, it may be interpreted in four different ways: (i) the wholesale destruction of the planet and the race, which will certainly occur if all the human eggs remain confined to our single terrestrial basket when our Sun 'goes nova' five billion years hence; (ii) the loss of our planet to some catastrophe or another, but the survival of at least some elements of our race on other worlds; (iii) the obliteration of the human race but the survival of the planet, due perhaps to some virulent and inescapable disease; and (iv) the end of the world *as we know it*. It is on this final scenario that I will be focusing here, and the main thrust of this book will address global geophysical events that have the potential to deal our race and our global technological society, if not exactly a lethal blow, then one that will certainly put us on the canvas. In other words, natural catastrophes on scales mighty enough to bring to an end our familiar world. I will not concern myself with technological threats such as those raised by advances in artificial intelligence and robotics, genetic engineering, nano-technology, and increasingly energetic high-energy physics experiments. Neither will I address attempts by some of the human race to reduce its numbers through nuclear, biological, or chemical warfare. Instead I want to introduce you to some of the very worst that Nature can throw at us, either solely on its own account or—in the case of contemporary climate change—with our unthinking help.

While often benign and even supportive, Nature can be a terrible foe and mankind has fought a near-constant battle against the results of its capriciousness—severe floods and storms, devastating earthquakes, and cataclysmic volcanic eruptions. The

Asian tsunami of 26 December 2004 provided us with just a taster of the worst Nature can do, destroying 400,000 buildings, killing 230,000 citizens from 40 countries—including up to 100,000 children—and leaving an astonishing eight million people homeless, unemployed, and impoverished. While the scale and extent of the tsunami's awful legacy is unprecedented in modern times, we have—on the whole—been quite fortunate, and our civilization has grown and developed against the tranquil background of the Holocene interglacial that has reigned now for more than 10,000 years. The omens for the next century and beyond are, however, dire. The rises in temperature and sea level that will characterize coming decades—in combination with ever-growing populations—will without doubt result in a huge increase in the number and intensity of natural disasters. Counter-intuitively, some parts of the planet may even end up getting chillier and it is at least feasible that the UK, for example, could—in this century—experience cooler conditions as Arctic ice melts, the Jet Stream is modified, and the Gulf Stream weakens. And what exactly happened to the predicted new Ice Age? Has the threat gone away with the onset of anthropogenic (man-made) global warming or are the glaciers simply biding their time?

While extraordinarily rapid in geological terms, climate change is a slow-onset event in comparison with the average human lifespan, and to some extent at least its progress can be measured and forecast. Much more unexpected and difficult to predict are those spontaneous geological events big enough and violent enough to lay waste to our entire society and which we have yet to experience in modern times. These can broadly be divided into extraterrestrial and terrestrial phenomena. The former involve the widely publicized threat to the planet arising from collisions with large chunks of space debris in the form of comets or asteroids. Even a relatively small, 2km object striking the planet could bring drastic changes to the global climate sufficient to wipe out perhaps a quarter of the human race.

The potential for the Earth itself to do us serious harm is less widely documented, but the threat of a global natural catastrophe arising from the bubbling and creaking crust beneath our feet is a real and serious one. Three epic events await us that have occurred many times before in our planet's prehistory, but which we have yet to experience in historic time. A cataclysmic volcanic 'super-eruption' plunged the planet into a bitter 'volcanic winter' some 74,000 years ago, while little more than 100,000 years ago gigantic waves caused by a collapsing Hawaiian volcano mercilessly pounded the entire coastline of the Pacific Ocean. Barely a thousand years before the birth of Christ, and again during the Dark Ages, much of eastern Europe and the Middle East was battered by an earthquake 'storm' that levelled once great cities over an enormous area. There is no question that such tectonic catastrophes will strike again in our future, but just what will be their effect on our global, technology-based society? How well we will cope is difficult to predict, but there can be little doubt that for most of the inhabitants of Earth, things will take a turn for the worse.

Living on the most active body in the solar system, we must always keep in our minds that we exist and thrive only by geological accident. As I shall address in Chapter 4, studies on human DNA have revealed that our race may have come within a hair's breadth of extinction following the unprecedented super-eruption 74,000 years before present, and if we had been around 65 million years ago when a 10km asteroid struck the planet we would have vanished alongside the dinosaurs. We must face the fact that, as long as we are all confined to a single planet in a single star system, prospects for the long-term survival of our race are always going to be tenuous. However powerful our technologies become, as long as we remain in Earth's cradle we will always be dangerously exposed to Nature's every violent whim. Even if we reject the 'doom soon' scenario, it is likely that our progress as a race will be continually impeded or knocked back by a succession of global natural catastrophes that will crop up at irregular intervals as long as the Earth exists and we upon it. While some of these events may bring to an end the world as we know it, barring another major asteroid or comet impact on the scale of the one that saw off the dinosaurs, the race itself is likely to survive.

The big question is, will we continue to advance technologically, or enter a period of slow decline culminating in eventual extinction? Optimists and technophiles are invariably convinced that the future is rosy and that at some point our race will begin to move out into space—first to our sibling worlds and then to the stars. Others—some might justifiably call them the realists—fail to see how our society's increasingly inward-looking focus could ever underpin a serious, concerted, move beyond our atmosphere. It seems, then, that we are close to a fork in the road. Either we are on the verge of the greatest of adventures that will see the human race breathe a collective sigh of relief as it guarantees its survival by installing proverbial eggs in baskets across the solar system and, ultimately perhaps, across the galaxy; or, we remain confined to our single, Earthbound nest, becoming ever more fragile and increasingly prone to a killer blow, whatever form that might take.

Which route we will follow is anyone's guess. As this book will show, when it comes to geophysics, what will be, will be.

Bill McGuire
Brassington, Derbyshire, England
October 2013

List of illustrations

The publisher and the author apologize for any errors or omissions in the above list. If contacted they will be pleased to rectify these at the earliest opportunity.

Chapter 1
Planet Earth: in a nutshell

Danger: Nature at work

We are so used to seeing on our television screens the pathetic remains of cities battered by earthquakes, or the thousands of terrified refugees fleeing the aftermath of yet another surging tsunami, that they no longer hold any surprise or fear for us, insulated as we are by distance and a lack of true empathy. Although not entirely immune to disaster themselves, the great majority of citizens fortunate enough to live in prosperous Europe, North America, or Oceania view great natural catastrophes as ephemeral events that occur in strange lands far, far away. Mildly interesting but only rarely impinging upon a daily existence within which a murder in a popular soap opera or a win by the local football team holds far more interest than six million people displaced by biblical floods across Pakistan. Remarkably, such an attitude even prevails in regions of developed countries, where Nature has previously spawned catastrophe but has been lying low for some time. Talk to the citizens of Naples in Italy about the threat of their local volcano exploding into life, or to the inhabitants of Paris, France, about prospects for their city being inundated by the Seine bursting its banks, and they are likely to shrug and point out that they have far more immediate things to worry about. The only explanation is that these people are in denial. They are quite aware that terrible disaster *will* strike at

1

some point in the future—they just can't accept that it might happen to them or their descendants.

When it comes to natural catastrophes on a global scale such an attitude is virtually omnipresent; pervading national governments, international agencies, multinational trading blocks, and much of the scientific community. There is some cause for optimism, however, and in one area, at least, this has begun to change. The threat to the Earth from asteroid and comet impacts is now common knowledge and the race is on to identify all those Earth-approaching asteroids that have the potential to stop the development of our race in its tracks. Thanks to widely publicized television documentaries and dramas shown in the UK and United States over the last couple of decades, the added threats of volcanic super-eruptions and giant tsunamis have now also begun to reach an audience wider than the tight groups of scientists that work on these rather esoteric phenomena. In particular, the blanket media coverage of the December 2004 Indian Ocean tsunami ensured that the phenomenon and its capacity for widespread devastation and loss of life is now understood and appreciated far and wide, even if this failed to prevent the loss of a further 16,000 lives when a comparable wave struck the north-east coast of Japan in March 2011.

To all who have paused to consider, it must be apparent that the Earth is an extraordinarily fragile place that is fraught with danger: a tiny rock hurtling through space, racked by violent movements of its crust and subject to dramatic climatic changes as its geophysical and orbital circumstances vary and its apex species does its worst. Barely 10,000 years after the end of the Ice Age, the planet is sweltering in some of the highest temperatures in recent Earth history. At the same time, overpopulation, resource depletion, and environmental degradation are dramatically increasing the vulnerability of modern society to natural catastrophes such as earthquakes, tsunamis, floods, and volcanic eruptions.

For better and for worse, the Earth is the most dynamic planet in our solar system; a dynamism that has given us our protective magnetic field, our atmosphere, our oceans, and ultimately our lives. The very same geophysical features that make our world so life-giving and life-preserving also, however, make it dangerous. The spectacular volcanoes that in the early history of our planet helped to generate the atmosphere and the oceans have, in the last three centuries, for example, wiped out more than a quarter of a million people and injured countless others. At the same time, the rains that feed our rivers and provide us with the potable water that we need to survive have inundated enormous tracts of the planet with floods that in recent years have been truly biblical in scale. On average, in any single year, floods impinge upon the lives of half a billion people and result in 25,000 deaths. Since 2010 alone, devastating floods have affected countries as physically distinct and geographically separated as Pakistan, India, China, Thailand, Japan, Australia, Canada, Brazil, Colombia, Germany, and the UK, bringing misery to billions of people. The lesson is, then, that Nature provides us with all our needs but we must be very wary of its rapidly changing moods.

The Earth: a potted biography

The major global geophysical catastrophes that surely await us down the line are run-of-the-mill natural phenomena writ large. In order to understand them, therefore, it is essential to know a little about the Earth and how it functions. To begin, it is worth considering just how astonishingly old the Earth is, if only to appreciate the notion that just because we have not experienced a particular natural catastrophe before does not mean it has never happened, nor that it will not happen again. Our world has been around for 4.6 billion years; just about long enough to ensure that anything Nature can conjure up it already has. Since the first single-celled organisms bloomed, billions of years ago, within sweltering chemical soups brooded over by a noxious atmosphere, life has struggled desperately to survive and evolve against a

3

background of potentially lethal geophysical phenomena. Little has changed today, except perhaps for a fall in the frequency of global catastrophes, and many on the planet still face a daily threat to life, limb, and livelihood from volcano, quake, flood, and storm. The natural perils that have assaulted our race in the past, and that constitute a growing future threat, have roots that extend back to the creation of the solar system and the formation of the Earth from a disc of debris orbiting a primordial Sun. Like our sister planets, the Earth can be viewed as a lottery jackpot winner; one of only eight chunks of space debris (nine if you are of a mind to count Pluto) out of original trillions that managed to grow and endure while the rest annihilated one another in spectacular collisions or were swept up by the larger lucky few with their stronger and more influential gravity fields. This sweeping-up process—known as accretion—involved the Earth and other planets adding to their masses through collisions with other smaller chunks of rock, an extremely violent process that was mostly completed—fortunately for us—close to four billion years ago. After this time, the solar system was a much less cluttered place, with considerably less debris hurtling about and impacts on the planets less ubiquitous events. Nevertheless, major collisions between the Earth and asteroids and comets—respectively rocky and rock-ice bodies that survived the enthusiastic spring cleaning during the early history of the solar system—are recognized throughout our planet's geological record. As I shall discuss in Chapter 5, such collisions have been held responsible for a number of mass extinctions over the past half a billion years, including that which doomed the dinosaurs. Furthermore, the threat of asteroid and comet impacts is still very much with us, and 1,429 Potentially Hazardous Asteroids (or PHAs) have already been identified whose orbits around the Sun bring them a little too close to the Earth for comfort.

The primordial Earth bore considerably more resemblance to our worst vision of hell than today's stunning blue planet. The enormous heat generated by collisions, together with that

4

produced by high concentrations of radioactive elements within the Earth, ensured that the entire surface was covered with a churning magma ocean, perhaps 400km deep. Temperatures at this time were comparable with some of the cooler stars, approaching 5,000°C. Inevitably, where molten rock met the bitter cold of space, heat was lost rapidly, allowing the outermost levels of the magma ocean to solidify to a thin crust. Even before its hundred millionth birthday, however, the growing geological calm was shattered by the unimaginably violent coming together of the Earth and a Mars-sized object; the resulting titanic collision gouging out a great chunk of our young world, forming a ring of debris that rapidly coalesced to form the Moon. A thin crust began to form again as things settled down, although the continuously churning currents in the molten region immediately below repeatedly caused this to break into fragments and slide once again into the maelstrom. By about 2.7 billion years ago more stable and long-lived crust managed to develop and to gradually thicken. Convection currents continued to stir in the hot and partially molten rock below, carrying out the essential business of transferring the heat from radioactive sources in the planet's deep interior into the growing rigid outer shell from where it was radiated into space. The disruptive action of these currents ensured that the Earth's rigid outer layer was never a single, unbroken carapace, but instead comprised separate rocky *plates* that moved relative to one another on the backs of the sluggish convection currents.

As a crust was forming, major changes were also occurring deep within the Earth's interior. Here, heavier elements—mainly iron and nickel—were slowly sinking under gravity towards the centre to form the planet's metallic core. At its heart, a ball made up largely of solid iron and nickel formed, but pressure and temperature conditions in the outer core were such that this remained molten. Being a liquid, this also rotated in sympathy with the Earth's rotation, in the process generating a magnetic field that later protected developing life on the surface by blocking

5

damaging radiation from space and ultimately provided us with a reliable means of navigation without which our pioneering ancestors would have found exploration—and returning home again—a much trickier business.

During the last couple of billion years or so, things quietened down considerably on the planet, and its structure and the geophysical processes that operate both within and at the surface have not changed a great deal. Internally, the Earth has a threefold structure. A crust made up of low-density, mainly silicate, minerals incorporated into rocks formed by volcanic action, sedimentation, and burial; a partly molten mantle consisting of higher-density minerals, also silicates; and a composite core of iron and nickel with some impurities. Ultimately, the hazards that constantly impinge upon our society result from our planet's need to rid itself of the heat that is constantly generated in the interior by the decay of radioactive elements. As in the Earth's early history, this is carried towards the surface by convection currents within the mantle. These currents in turn constitute the engines that drive the great, rocky plates across the surface of the planet, and underpin the concept of plate tectonics, which geophysicists use to provide a framework for how the Earth operates geologically.

The relative movements of the plates themselves, which comprise the crust and the uppermost rigid part of the mantle (together known as the lithosphere), are in turn directly related to the principal geological hazards—earthquakes and volcanoes, which are concentrated primarily along plate margins (see Figure 1). Here a number of interactions are possible. Two plates may scrape jerkily past one another, accumulating strain and releasing it periodically through destructive earthquakes. Examples of such conservative plate margins include the quake-prone San Andreas Fault that separates western California from the rest of the United States and Turkey's North Anatolian Fault, whose latest movement triggered a devastating earthquake in 1999, and whose next is awaited—with some trepidation—by the inhabitants of

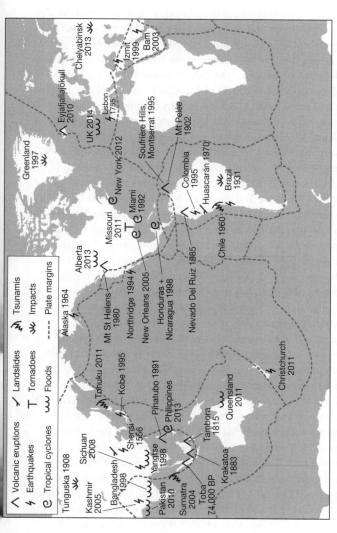

1. **Map of the Earth's plates with locations of recent natural disasters: the locations of many natural disasters coincide with the plate margins**

7

Istanbul. Alternatively, two plates may collide head on. If they both carry continents built from low-density granite rock, as with the Indian Ocean and Eurasian plates, then the result of collision is the growth of a high mountain range—in this case the Himalayas—and at the same time the generation of major quakes such as that which wiped out 75,000 lives in Pakistan-administered Kashmir in October 2005. On the other hand, if an oceanic plate made of dense basalt hits a low-density continental plate then the former will plunge underneath its lighter companion, pushing back into the hot, convecting mantle. As one plate thrusts itself beneath the other (a process known as subduction) so the world's greatest earthquakes are generated. These include huge earthquakes in Chile in 1960, Alaska in 1964, and—most recently—Sumatra (Indonesia) in 2004 and northern Japan in 2011; all four triggered devastating tsunamis. Subduction is going on all around the Pacific Rim, ensuring high levels of seismic activity in Alaska, Japan, Taiwan, the Philippines, Chile, and elsewhere in the circum-Pacific region. This type of destructive plate margin—so called because one of the two colliding plates is destroyed—also hosts large numbers of active volcanoes. Although the mechanics of magma formation in such regions is sometimes complex, it is ultimately a result of the subduction process and owes much to the partial melting of the subducting plate as it is pushed down into ever hotter levels in the mantle. Fresh magma formed in this way rises as a result of its low density relative to the surrounding rocks, and blasts its way through the surface at volcanoes that are typically explosive and particularly hazardous. Strings of hundreds of active and dormant volcanoes circle the Pacific, making up the legendary Ring of Fire, while others sit above subduction zones in the Caribbean and Indonesia. Virtually all large, lethal eruptions occur in these areas, including those contributing to volcanic disasters at Pinatubo (Philippines) in 1991, Rabaul (Papua New Guinea) in 1994, and Montserrat (Lesser Antilles, Caribbean); in the latter case an eruption which—at the time of writing—is rumbling on into its 19th year.

To compensate for the consumption of some plate material, new rock must be created to take its place. This happens at so-called constructive plate margins, along which fresh magma rises from the mantle, squeezes into the gap between the plates on either side, and solidifies. This occurs beneath the oceans along a 40,000km long network of linear topographic highs known as the Mid-Ocean Ridge system, where newly created lithosphere exactly balances that which is lost back into the mantle at destructive margins (see Figure 2). A major part of the Mid-Ocean Ridge system runs down the middle of the Atlantic Ocean, bisecting Iceland, and separating the Eurasian and African plates in the east from the North American and South American plates in the west. Here too there are both volcanoes and earthquakes, but— notwithstanding the 2010 aviation-disrupting blast of Eyjafjallajökull—the former tend to involve relatively mild eruptions and the latter are small. Driven by the mantle convection currents beneath, the plates waltz endlessly across the surface of the Earth, at about the same rate as fingernails grow, constantly modifying the appearance of our planet and ensuring that, throughout the immensity of geological time, everywhere gets its fair share of earthquakes and volcanic eruptions.

Hazardous Earth

While earthquakes and volcanic eruptions are linked to how our planet functions geologically, other geophysical hazards are more dependent upon processes that operate in the Earth's atmosphere. Our planet's weather machine is driven by energy from the Sun, aided and abetted by the Earth's rotation and the constant exchange of energy and water with the oceans. As such, our local star is the ultimate progenitor of the tropical cyclones and floods that exact such an enormous toll on life and property, particularly in developing countries. Still other lethal natural phenomena have a composite origin and are less easy to pigeonhole. The giant sea waves known as tsunamis, for example, can be formed in a number of different ways; most commonly by submarine earthquakes, but

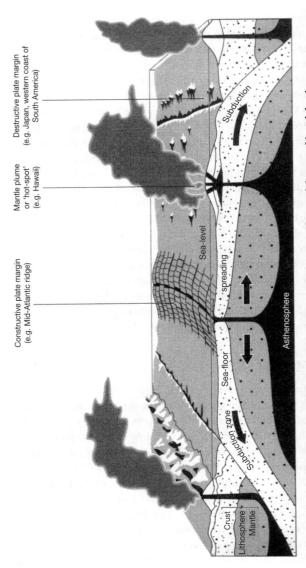

Destructive plate margin
(e.g. Japan, western coast of South America)

Mantle plume or 'hot-spot'
(e.g. Hawaii)

Constructive plate margin
(e.g. Mid-Atlantic ridge)

Subduction

Sea-level

spreading

Sea-floor

Subduction zone

Asthenosphere

Crust
Lithosphere
Mantle

2. The lithosphere, the Earth's outer rigid shell, is created at mid-ocean ridges and destroyed in subduction zones

10

also by landslides into the ocean and by eruptions of coastal and island volcanoes. Similarly, many landslides result from collusion between geology and meteorology, with torrential rainfall destabilizing already weak slopes.

At any single point, and at any one time, the Earth and its enclosing atmospheric envelope give the impression of being mundanely stable and benign. This is, however, an entirely misleading notion, with something like 3,500 detected earthquakes rocking the planet every day and a volcano erupting every week. Each year, warmer climes are battered by up to 40 tropical cyclones, the latest of which—Super-typhoon Haiyan— took more than 6,000 lives across the Philippines in 2013, while floods and landslides occur everywhere in numbers too great to keep track of.

In terms of the number of people affected—at least 100 million people a year—floods undoubtedly constitute the greatest of all natural hazards, a situation that is likely to continue given a future of rising sea levels and more extreme precipitation. River floods are respecters of neither wealth nor status, and both developed and developing countries have been severely afflicted in recent years, across every continent. Wherever rain is unusually torrential or persistent, it will not be long before river catchments fail to contain surface run-off and start to expand across their flood plains and beyond. In fact, the intensity of rainfall can be quite astonishing, with, in 1970, nearly 4cm of rain falling in just 60 *seconds* on the French Caribbean island of Guadeloupe—a world record. On another French island, Réunion, in the Indian Ocean, a passing cyclone dropped getting on for 2 *metres* of rain during a single 24-hour period in January 1966. As flood plains all over the world become more crowded, the loss of life and damage to property caused by swollen rivers has increased dramatically. In the spring of 1993, the Mississippi and Missouri rivers burst their banks, inundating nine Midwest states, destroying 50,000 homes, and leaving damage totalling US$20 billion. Massive floods

11

occurred in many parts of the UK in autumn 2000 as rain fell with a ferocity not seen for over 300 years, and again during the winter of 2013. Much of central Europe was swamped beneath record flood waters in the summer of 2002; an event almost matched in scale by massive flooding across the region in 2013. In 2010, monsoon floodwaters extended across a staggering one-fifth of Pakistan's land area, while three years later, northern India bore the brunt of flash floods reported to have taken close to 6,000 lives. Much of the Australian state of Queensland was under water in January 2011, while in June 2013, it was the turn of the Canadian province of Alberta to face catastrophic flooding. Across China, flooding on the Yangtze, Yellow, and other rivers continues to present a major threat, and is held responsible for over five million deaths over the last century and a half. Bangladesh has it even worse, the country often finding two-thirds of its land area under water as a result either of floodwaters pouring down the great Ganges river system or of cyclone-related storm surges hurtling inland from the Bay of Bengal. Coastal flooding due to storms probably takes more lives than any other natural hazard, with an estimated 300,000 losing their lives in Bangladesh in 1970 and 15,000 in Orissa, north-east India, in 1999.

Partly through their effectiveness at spawning floods, but also as a consequence of the enormous wind speeds achieved, storms constitute one of the most destructive of all natural hazards. Furthermore, because they are particularly common in some of the world's most affluent regions, they are responsible for some of the most costly natural disasters of all time. Every year, the Caribbean, the Gulf and southern states of the USA, and Japan are struck by tropical cyclones, while the UK and continental Europe suffer from severe and damaging winter storms. In 1992, Hurricane Andrew virtually obliterated southern Miami in one of the costliest natural disasters in US history, resulting in losses of US$32 billion. This epic storm brought to bear on the city wind speeds of up to 300km per second, leaving 300,000 buildings damaged or destroyed and 150,000 homeless. The 2004 and 2005

Atlantic hurricane seasons were extraordinarily active, the state of Florida being struck by six hurricanes over a 12-month period. The 2005 season also spawned Hurricane Katrina, which far exceeded Andrew in its destructiveness, killing close to 2,000 people and leaving 80 per cent of the city of New Orleans awash in up to 7m of flood water. Destructive windstorms are not only confined to the tropics, and hurricane-force winds also accompany low-pressure weather systems at mid-latitudes. Many residents of southern England will remember the Great Storm of October 1987 that felled millions of trees with winds whose average speeds were clocked at just below hurricane force. In 1999, France suffered a similar ordeal as winter storm Lothar blasted its way across the north of the country. On the other side of the 'pond' the US Midwest braces itself every year for a savage onslaught from tornadoes: rotating maelstroms of solid wind that form during thunderstorms in the contact zone between cold, dry air from the north and warm, moist air from the tropics. No man-made structures that suffer a direct hit can withstand the average wind speeds of up to 500km an hour, and damage along a tornado track is usually total. Although rarely as lethal as hurricanes, in just a few days in April 1974 almost 150 tornadoes claimed more than 300 lives in Kentucky, Tennessee, Alabama, and adjacent states. Almost equally lethal, in May 2011 a single monster tornado all but obliterated the Missouri town of Joplin, killing 158 of its residents.

Of the so-called geological hazards—earthquakes, volcanic eruptions, and landslides—there is no question that earthquakes are by far the most devastating. Every year about 150 quakes reach or exceed magnitude six on the well-known Richter Scale, making them large enough to cause significant damage and loss of life, particularly when they strike poorly constructed and ill-prepared population centres in developing countries. As previously mentioned, most large earthquakes are confined to distinct zones that coincide with the margins of plates. In the last few decades, sudden movements of California's San Andreas Fault have

13

generated large earthquakes in San Francisco (1989) and southern California (1994), the latter causing damage amounting to US$35 billion—at the time the costliest natural disaster in US history. Just a year later, a magnitude 7.2 quake at the western margin of the Pacific plate devastated the Japanese city of Kobe, killing 6,000 and engendering economic losses totalling a staggering US$150 billion; making the event, until the 2011 Tohuku quake and tsunami, the most expensive natural disaster of all time. Four years after Kobe, the North Anatolian Fault ruptured just to the east of Istanbul, triggering a severe quake that flattened the town of Izmit and neighbouring communities and took over 17,000 lives. Since the end of the millennium, massive quakes at or close to plate margins have shaken—amongst other nations—Iran, Indonesia, Pakistan, Chile, Haiti, Japan, China, and New Zealand, together taking more than 700,000 lives, including those lost to associated tsunamis. Large earthquakes can also occur, however, at locations remote from plate margins, and have been known in northern Europe and the eastern USA, which are not regions of high seismic risk. One such intraplate quake devastated the Bhuj region of India's Gujarat state in January 2001, completely destroying 400,000 buildings and killing perhaps as many as 100,000 people. Another shook L'Aquila in central Italy in 2009, taking close to 300 lives and prompting a high-profile inquest into the role of Italian seismologists and government officials in the run-up to the quake. This culminated in six individuals being found guilty of multiple manslaughter for playing down the likelihood of a major quake a few days before it took place.

There is a truism uttered by earthquake engineers: *earthquakes don't kill people; buildings kill people*. Without question this is the case, and both damage to property and loss of life could be drastically reduced if appropriate building codes were both applied and enforced. Earthquakes also, however, prove lethal through the triggering of landslides as a result of ground shaking, and by the formation of tsunamis. The latter are generated when a quake instantaneously jerks upwards—perhaps by just a metre or

14

3. The devastating aftermath of the 2011 tsunami at Kesennuma in the prefecture of Miyagi in north-east Japan

so—a large area of the seabed, causing the displaced water above to hurtle outwards as a series of waves. When these enter shallow water they build in height—sometimes to 30m or more—and crash into coastal zones with extreme force. Together, the 2004 Indian Ocean tsunami and the Japan tsunami of 2011 took close to a quarter of a million lives and destroyed or damaged millions of homes and other buildings (see Figure 3).

Estimates of the number of active volcanoes vary, but there are at least 1,500 and possibly more than 3,000. Every year around 50 volcanoes erupt, some of which—like Kilauea on Hawaii or Stromboli in Italy—are almost constantly active. Others, however, may have been quiet for centuries or in some cases millennia and these tend to be the most destructive. The most violent volcanoes occur at destructive plate margins, where one plate is consuming another. Their outbursts rarely produce quiet flows of red lava and are more likely to launch enormous columns of ash and debris 20km or more into the atmosphere. Carried by the wind over huge areas, volcanic ash can be extremely disruptive, making travel

15

difficult, damaging crops, poisoning livestock, and contaminating water supplies. Just 30cm or so of wet ash is sufficient to cause roofs to collapse, while the fine component of dry ash can cause respiratory problems and illnesses such as silicosis. Close to an erupting volcano the depth of accumulated ash can reach several metres, sufficient to bury single-storey structures. This was the fate of much of the town of Rabaul on the island of New Britain (Papua New Guinea), during the 1994 eruptions of its twin volcanoes Vulcan and Tavurvur. For years following the 1991 eruption of Pinatubo in the Philippines, thick deposits of volcanic debris provided a source for mudflows whenever a tropical cyclone passed overhead and dumped its load of rain. Almost a decade later, mud pouring off the volcano was still clogging rivers, inundating towns and agricultural land, and damaging fisheries and coral reefs. Somewhat surprisingly, mudflows also constitute one of the biggest killers at active volcanoes. In 1985 a small eruption through the ice and snow fields of Columbia's Nevado del Ruiz volcano unleashed a torrent of mud out of all proportion to the size of the eruption, which poured down the valleys draining the volcano and buried the town of Armero along with 23,000 of its inhabitants.

Even scarier and more destructive than volcanic mudflows are pyroclastic flows or 'glowing avalanches'. These hurricane-force blasts of incandescent gas, molten lava fragments, and blocks and boulders sometimes as large as houses have the power to obliterate everything in their paths. In 1902, in the worst volcanic disaster of the 20th century, pyroclastic flows from the Mont Pelée volcano on the Caribbean island of Martinique annihilated the town of St Pierre as effectively as a nuclear bomb; within a few minutes leaving only two survivors out of a population of close to 29,000 (see Figure 4). The threat from volcanoes does not end there: chunks of rock collapsing into the sea from their flanks can trigger colossal tsunamis, while noxious fumes can and have killed thousands and their livestock. Volcanic gases carried into the stratosphere, and from there around the planet, have modified the

4. The ruins of St Pierre (Martinique) after 1902 eruption: only two inhabitants of St Pierre survived the onslaught of the Mont Pelée volcano

climate and led to miserable weather, crop failures, and health problems half a world away. On the grandest scale, volcanic super-eruptions have the potential to affect us all, through plunging the planet into a frigid 'volcanic winter' and devastating harvests worldwide.

Of all geological hazards, landslides are perhaps the most underestimated, probably because they are often triggered by some other hazard, such as an earthquake or deluge, and the resulting damage and loss of life is therefore subsumed within the tally of the primary event. Nevertheless, landslides can be highly destructive, both in isolation and in numbers. In 1556, a massive earthquake struck the Chinese province of Shaanxi, shaking the ground so vigorously that the roofs of countless cave dwellings collapsed, burying alive (according to Imperial records) over 800,000 people. In 1970, another quake caused the entire peak of the Nevados Huascaran mountain in the Peruvian Andes to fall on the towns below, snuffing out the lives of 18,000 people in just four minutes and erasing all signs of their existence from the face of the Earth. Heavy rainfall too can be particularly effective at

17

triggering landslides and when, in 1998, Hurricane Mitch dumped up to 60 centimetres of rain on Central America in 36 hours, it mobilized more than a million landslides in Honduras alone, blocking roads, burying farmland, and destroying communities.

The final—and perhaps greatest—threat to life and limb comes not from within the Earth but from without. Although the near constant bombardment of our planet by large chunks of space debris ended billennia ago, the threat from asteroids and comets remains real and is treated increasingly seriously. Recent estimates suggest that around a thousand asteroids with diameters of 1km or more have orbits around the Sun that approach or cross the Earth's, making collision possible at some point in the future; this population includes many objects 2km across and larger. An object this big striking our planet would trigger a 'cosmic winter' due to dust lifted into the stratosphere blocking out solar radiation, perhaps wiping out a quarter or so of the human population as a result. The revival of interest in the impact threat arose largely as a consequence of two important scientific events that occurred in the last decade of the 20th century: first, the identification of a large impact crater at Chicxulub, off Mexico's Yucatan Peninsula, which is now widely regarded as the 'smoking gun' responsible for global genocide at the end of the Cretaceous period; second, the eye-opening collisions, in 1994, of the fragments of Comet Shoemaker-Levy with Jupiter. Images flashed around the world of resulting impact scars larger than our own planet were disconcerting to say the least and begged the question in many quarters—what if that were the Earth?

Natural hazards and us

If you were not already aware of the scale of the everyday threat from Nature then I hope, by now, to have engendered a healthy respect for the destructive potential of the hazards that many of our fellow inhabitants of planet Earth have to face almost on a

18

daily basis. The reinsurance company Munich Re., who, for obvious reasons, have a considerable interest in this sort of thing, estimate (rather conservatively I suspect) that up to 15 million people were killed by natural hazards in the last millennium, and over 3.5 million in the last century alone. The most lethal year was 2004, during which the Indian Ocean tsunami, together with earthquakes in Morocco and Japan, record storms in the USA and Japan, and flooding across Asia, contributed to one-third of a million deaths. In 2011, however, the economic cost of natural hazards to the global economy reached unprecedented levels, with earthquakes in Turkey and New Zealand, severe floods in Thailand, Australia, China, and Brazil, a massive tornado outbreak in the United States, and storms in Korea, Japan, the Philippines, the Caribbean, and western Europe—along with the Japan tsunami—making up the bulk of economic losses that totalled a colossal US$380 billion.

The first decade of the 20th century saw more than 2.5 billion people suffer due to natural disasters. Unhappily, there is little sign that hazard impacts on society are diminishing as a consequence of improvements in forecasting and hazard mitigation, and the outcome of the battle against Nature's dark side remains far from a foregone conclusion. While we now know far more about natural hazards, the mechanisms that drive them, and their sometimes awful consequences, any benefits accruing from this knowledge have been at least partly negated by the increased vulnerability of large sections of the Earth's population. This has arisen primarily as a result of the rapid rise in the number of people, which doubled between 1960 and 2000 and, in 2011, topped the seven billion mark. The bulk of this rise has occurred in poor developing countries, many of which are particularly susceptible to a whole spectrum of natural hazards. Furthermore, the struggle for living space has ensured that marginal land, such as steep hillsides, flood plains, and coastal zones, has become increasingly utilized for farming and habitation. Such terrains are clearly high risk and can expect to

19

succumb on a more frequent basis to, respectively, landsliding, flooding, storm surges, and tsunamis.

Another major factor in raising vulnerability in recent years has been the move towards urbanization in the most hazard-prone regions of the developing world. In 2007, for the first time ever, the number of people living in urban environments overtook the number residing in the countryside, many crammed into poorly sited and badly constructed megacities with populations in excess of 8 million people. Forty years ago New York and London topped the league table of world cities, with populations, respectively, of 12 and 8.7 million. Today, however, cities such as Jakarta (Indonesia), Karachi (Pakistan), and Mexico City are firmly ensconced in the top ten (see Table 1); gigantic sprawling agglomerations of humanity with populations approaching or exceeding 20 million, and extremely vulnerable to storm, flood, and quake. A staggering 96 per cent of all deaths arising from

Table 1 The ten most populous cities in the world in 2013

Rank	City	Population (million)
1	Tokyo (Japan)	37.1
2	Jakarta (Indonesia)	26
3	Seoul (South Korea)	22.5
4	Delhi (India)	22.2
5	Shanghai (China)	20.8
6	Manila (Philippines)	20.7
7	Karachi (Pakistan)	20.7
8	New York (USA)	20.4
9	Sao Paolo (Brazil)	20.1
10	Mexico City (Mexico)	19.4

Source: United Nations

natural hazards and environmental degradation occur in developing countries and there is currently no prospect of this falling. Indeed, the picture looks as if it might well deteriorate even further. With so many people shoehorned into ramshackle and dangerously exposed cities (many in coastal locations at risk from earthquakes, tsunamis, windstorms. and coastal floods) it can only be a matter of time before we see the first of a series of true mega-disasters, with death tolls exceeding one million.

The picture I have painted is certainly bleak, but the reality may be even worse. Future rises in population and increases in vulnerability will take place against a background of dramatic climate transformation, the like of which the planet has not experienced for more than 10,000 years. The jury remains out on the precise hazard implications of the rapid warming expected over the next hundred years, but rises in sea level that may exceed one metre are forecast in the most recent (2013) report of the IPCC (Intergovernmental Panel on Climate Change); a prediction that is regarded as optimistic by a number of prominent climate scientists. This will certainly increase the incidence and impact of storm surges and tsunamis and, in places, increase the degree of coastal erosion. Other consequences of a global average temperature rise that could reach close to 5°C by the end of the century will be more extreme meteorological events such as powerful hurricanes, tornadoes, and floods, greater numbers of landslides in mountainous terrain, and maybe even more earthquakes and volcanic eruptions.

So is the world as we know it about to end and, if so, how? A century from now will we be gasping for water in an increasingly roasting world or huddling around a few burning sticks, struggling to keep at bay the bitter cold of a cosmic winter? In the next chapter I shall delve a little further into the possibilities.

Chapter 2

Global warming: a lot of hot air?

Debate—what debate?

There is absolutely no question that the Earth is warming up fast, and virtually no reputable climate scientist would argue with this. The 'dispute' manufactured by the climate change deniers is rooted in whether or not the warming we are now experiencing simply reflects a natural turnabout in the recent global temperature trend or results from the polluting impact of human activities since the industrial revolution really began to take hold. Forecasting climate change is extremely difficult, which explains why models for future temperature rise and sea-level change are constantly undergoing revision, but the evidence is now irrefutable: human activities *are* driving the current period of planetary warming.

Notwithstanding a few maverick scientists, oil company apologists, dodgy television presenters, and over-the-hill politicians, the overwhelming consensus amongst those who have a grasp of the facts is that without a massive reduction in greenhouse gas emissions things are going to get very bad indeed. To our children's and their children's detriment, however, this unequivocal prospect is continually played down and intentionally hidden behind a veil of obfuscation by those whose neoliberal world view cannot and will not accept any scenario—true or not—that threatens to intrude upon and rein in an unrestrained capitalist free market. Unfortunately, the climate does not recognize political stance or ideological rhetoric, so the world will continue to heat up, whatever half-baked nonsense the climate change deniers continue to spout.

During the past 80 years, the Earth has been hotter than at any other time in the last millennium, and the warming has accelerated dramatically in just the past few decades. In fact, of the 14 hottest years ever recorded, 12 occurred during the period 2001–12. Hardly surprising, then, that the decade 2000–9 is the hottest on record, followed by the 1990s and the 1980s. There may be a few ups and downs as far as year-on-year temperatures are concerned, due to ephemeral climate influences such as El Niño, but the overall trend is remorselessly upwards. In its 2013 5th Assessment Report, the IPCC records that global average temperatures rose by 0.8°C between 1901 and 2010, with more than half of this rise coming in the 31 years between 1979 and 2010. The Earth is now warmer than it has been for over 90 per cent of its 4.6 billion year history, and by the end of the 21st century our planet may see higher temperatures than at any time for the last 150,000 years. The hard truth is that our world is now heating up at a faster rate than at any time in the last 11,500 years, and there is no sign of any slowdown (see Figure 5).

The rising temperature trend we are seeing now is not simply a climatic blip or hiccough, nor can it be explained entirely, as some would still have it, by variations in the output of the Sun. Although the output of our local star clearly does have a significant effect on the climate, there has been no concomitant increase in the Sun's output as global temperatures have risen over the course of the last 100 years or so. Furthermore, the Sun's activity has recently been at its lowest level for a century and there have even been suggestions that it might be entering a

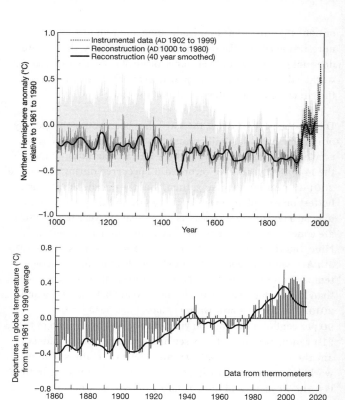

0.8

Departures in global temperature (°C) from the 1961 to 1990 average

0.4

0.0

-0.4

Data from thermometers

-0.8

1860 1880 1900 1920 1940 1960 1980 2000 2020
Year

5. Northern hemisphere temperature change since 1000AD (top) and global temperature change since 1860 (bottom)

'grand minimum': a state of reduced activity that, last time it happened, contributed to the bitter cold of the Little Ice Age that extended from the late 15th century to the second half of the 19th. Rather, the current warming is a consequence of two centuries of pollution, which is now enclosing the Earth in an insulating blanket of carbon dioxide, methane, nitrous oxide, and other greenhouse gases. Since the late 18th century our race has been engaged in a gigantic planetary trial, the final outcome

24

of which we can still only guess at. Unfortunately for us the experiment has now entered a runaway phase, which, due to its inherent inertia, we cannot stop but only slow down. Even if we were to stabilize greenhouse gas emissions today, both temperatures and sea levels would continue to rise for many hundreds of years.

The great global warming experiment

We know from studies of polar ice cores that before the hiss of steam and grinding of metal on metal that heralded the arrival of the industrial world, the concentrations of greenhouse gases in the atmosphere had been pretty much constant since the glaciers retreated at the end of the last Ice Age. Since pre-industrial times, however, carbon dioxide levels in the atmosphere have risen by 43 per cent, alongside sharp increases in other greenhouse gases, in particular methane and nitrous oxide. In the summer of 2013, atmospheric concentrations of carbon dioxide exceeded the symbolic 400 parts per million (ppm) threshold, bringing levels to their highest for at least 15 *million* years (see Figure 6). The rate of increase in the gas has also been quite unprecedented, and was greater in the last hundred years than at any time in at least the previous 20,000. Being concoctions of the 20th century, other polluting gases such as chlorofluorocarbons and hydrofluorocarbons were not even present in the atmosphere a couple of centuries ago. As all these gases have accumulated in the Earth's atmosphere so they have, quite literally, caused it to act in the manner of a greenhouse, allowing heat from the Sun in but hindering its escape back out into space. In fact, our atmosphere has operated in this way for billions of years, moderating temperature swings and extremes, but our pollution is now strongly enhancing this greenhouse effect, with the result that the Earth has been progressively warming up for most of the last hundred years.

Because the climate machine is so complex, no single influence can be taken in isolation and many other factors affect global

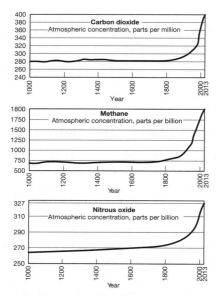

6. **Concentrations of carbon dioxide have risen dramatically since the industrial revolution and passed 400ppm in 2013. The levels of other greenhouse gases in the atmosphere also continue to climb**

temperatures. Not least of these is the output of the Sun, which is also variable over time, and which must be taken into account before allocating a rising temperature trend purely to the accumulation of anthropogenic greenhouse gases. The Sun follows a regular 11-year pattern of activity, known as the sunspot cycle, during which time its output varies by about 0.1 per cent. Solar output also changes over longer periods, ranging from hundreds to tens of thousands of years, and these can play a significant role in cooling or warming the planet and—in recent centuries—in modifying or masking the effect of gases arising from human activities. As I will address in more detail in Chapter 4, volcanic eruptions can also have a significant effect on the Earth's climate. Although the detailed picture is somewhat more complicated, large explosive eruptions inject massive volumes of sulphur dioxide and other sulphur gases into the stratosphere, which have a broadly cooling effect through reducing the level of solar radiation reaching the Earth's surface. Significant, if short-lived, reductions in global temperatures followed the eruptions of both El Chichón (Mexico) in 1982 and Pinatubo (Philippines) in 1991. Sometimes volcanoes and the Sun combine to bring about longer-lasting episodes of climate change. For example, a combination of reduced solar output and elevated volcanic activity has been implicated in the extended cold snap known as the Little Ice Age.

Attempting to pin down the true variation in global temperatures over the past thousand years is difficult, not least because records prior to the last couple of hundred years are far from reliable. A further complication arises from the fact that while one part of the world might be heating up, another might be cooling down. One argument that is still used by opponents of anthropogenic warming is that the world underwent a pronounced cooling between 1946 and 1975, thereby invalidating the idea that elevated levels of greenhouse gases must automatically result in global warming. More detailed examination of the record for this period reveals, however, that although much of the northern hemisphere cooled noticeably, the reverse was the case in the southern hemisphere, which warmed appreciably. In reality, although there was a small overall temperature fall at this time, this is now being attributed to a masking of the warming trend by sulphur gases emitted by volcanic eruptions and by heavy industries at the time unfettered by clean air laws. The bad news is that the diminution of this masking effect—now fashionably referred to as global dimming—as clean air legislation becomes more prevalent and clean energy sources are embraced may mean that temperatures are set to rise even further than forecast. In any case, notwithstanding this blip, instrumental records show that global temperatures have been following an inexorably upward path since such records began in 1861. On the Continent, the veracity of global warming really hit home during the summer of 2003, when record-breaking temperatures across the Continent claimed more than 70,000 lives due to heat stress, and during the even more extreme 2010 Russian heatwave, which resulted in at least 56,000 deaths.

If our great experiment was designed specifically to heat up the planet, then based upon the results to date it seems that we can pat one another on the back at a job well done and sit back and relax as the experiment grinds away of its own accord, racking up the heat and clocking up an ever-increasing list of unexpected consequences. But of course, this was not the intention of the experiment at all. Indeed, it is only in recent decades that the polluting effect of human activities on the global environment has been thought of in these terms. The great experiment has never been anything other than a by-blow of our race's constant thirst for more: more goods, more wealth, more growth. Now that it has become apparent that we have been messing, admittedly involuntarily, with the natural functioning of the Earth, we have no choice but to close the experiment down. So far, however, there has been little success. Continuing political procrastinations and the muddying of the scientific waters by vested interest groups antagonistic to proposals to mitigate global warming has ensured that, although eventually ratified, the first commitment period of the Kyoto Protocol fell far short of achieving its goal of a 5.2 per cent reduction (below 1990 levels) in global greenhouse gas emissions by 2008–12, and the second period has yet to get off the ground. There were great hopes for a breakthrough at the UN Climate Change Summit in Copenhagen in December 2009, with some hoping that a truly global agreement might come about, built around a level of emissions cuts that the science says is needed if dangerous, all-pervasive, climate change is to be avoided, but—as ever—the big players, China, the United States, and others, managed together to put a spanner in the works. True, a 'Copenhagen Accord' was agreed that acknowledged climate change as a great challenge and that actions should be taken to keep the global average temperature rise beneath 2°C, but the

28

absence of any legally binding commitment to emissions cuts meant that the Accord was close to worthless. Continued obfuscation and fiddling at more recent climate change summits would have made Emperor Nero truly proud. Looking the other way when your capital city is burning is one thing; doing so when the planet's climate is collapsing around you requires a degree of chutzpah on an altogether different scale.

Whatever measures we take now, it will still be a very long time before we see any response from the climate. Rather like trying to turn around a supertanker, the enormous inertia that has already built up in the climate system will ensure that dramatic changes to our environment cannot be avoided—a very few for the better, but most not. What is certain is that our children and their descendants are going to inhabit a planet that would be almost unrecognizable to us.

Hothouse Earth

The world of AD 2100 will not only be far warmer but will also be characterized by extremes of weather that will ensure, at the very least, a far more uncomfortable life for billions. Already, the wildly fluctuating weather patterns that are held by many to be a consequence of global warming, combined with increased vulnerability in the developing world, are leading to a dramatic rise in the numbers of meteorological disasters. According to the World Meteorological Organization, more than 370,000 people died in extreme weather incidents between 2001 and 2010; a 20 per cent rise over the previous decade. Few think that the situation will get better and it is virtually certain that things will get progressively worse.

Increasingly, those occupying low-lying coastal regions will be hit by rising sea levels that will mean that lethal floods become the norm rather than the exception. In contrast, more and more people will starve as annual rains fail year after year and huge regions of

Global warming: a lot of hot air?

29

Africa and Asia fall within the grip of drought and consequent famine. The events of 2011 provided a glimpse of what we can expect to become the norm, with record floods and landslides in Brazil taking 1,000 lives and destroying thousands of homes and unprecedented flooding inundating more than a million buildings in Thailand. At the same time, a staggering 13 million men, women, and children suffered drought and famine across north-east Africa, while severe drought conditions across the south-west United States triggered agricultural losses totalling US$8 billion. It also looks as if the Earth will become a windier place, with warmer seas triggering more powerful, and therefore potentially more destructive, storms, particularly in the tropics. I will return to the manifold hazard implications of global warming later, but let's look now at the latest predictions for temperature rise over the next 100 years. After all, this is the critical element that will drive the huge changes to our environment in this century and beyond.

In 2013, the IPCC published the first of its three 5th Assessment Reports on climate change, this one focusing on the physical science underpinning global warming. The IPCC was established in 1988 by the UN Environmental Programme and the World Meteorological Organization, with a remit to provide an authoritative consensus of scientific opinion on climate change using the best available expertise. The important word here is *consensus*. Approaching 1,000 scientists have been involved in putting together each of the assessments, the fourth—published in 2007—leaving little doubt of their validity except in the minds of the irrationally sceptical, the eternally optimistic, or the downright Machiavellian. Speak off the record to many climate scientists, however, and they will offer the view that the IPCC assessments are typically conservative in relation to the contemporary impact of climate change and optimistic about future prospects, and the Fifth Assessment is no exception in this regard. In particular, criticism this time focuses on the absence of any realistic evaluation of the consequences of feedback effects capable of dramatically increasing the rate and degree of warming; most notably the wholesale release of the potent greenhouse gas methane from thawing Arctic permafrost. So when the IPCC forecast that global average temperatures by 2016–35 could be between 0.4°C and 1.0°C higher than for the period 1986–2005, and up to nearly 5°C higher by 2081–2100, they could be seriously downplaying the actual level of warming. Even if our luck holds out and the Arctic does not start to offload its huge store of methane into the atmosphere, the IPCC's figures remain very scary. If 5°C does not sound like much, consider that a comparable rise transformed our ice-bound planet into the relatively balmy world we live in today; the transition involving enormous changes in the Earth's environment, not only in the climate and weather but also in vegetation and animal life. Furthermore, this figure is a global average, so that a 5°C rise across the planet would translate into much higher land temperatures and a rise well into double figures at high latitudes, threatening catastrophic melting of the polar ice sheets (see Figure 7).

There is every reason to expect that if and when the post-glacial temperature rise doubles again we will experience equally dramatic changes. This time, however, there are two important differences. First, the Earth has to feed, clothe, and support more than seven billion souls (projected to rise to more than nine billion by 2050), rather than a few million. Second, today's comparable temperature rise is taking place over the course of just a hundred years rather than thousands. Many of the consequences of such a stratospheric (excuse the pun) rise in global temperatures are obvious, but others less so. The polar regions and mountainous areas with permanent snow and ice are already suffering, and warming will continue to exact a severe toll here. Over the last 100 years there has been a massive retreat of mountain glaciers all over the world, while since the 1950s the Arctic sea ice has started to waste away at an astonishing rate, with the result that the North Pole was ice free in summer 2000. In recent years, the loss of the sea ice has reached unprecedented levels, with ice cover

31

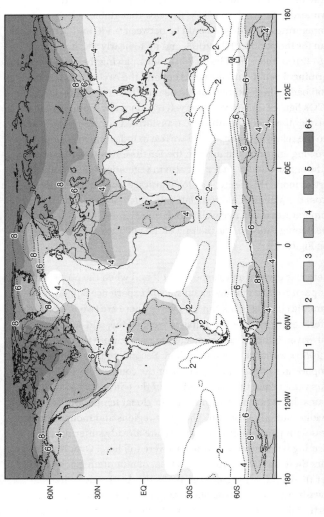

7. Projected annual mean change in temperature between the period 1961 to 1990 and 2071–2100, assuming a moderate rate of increase of greenhouse gas emissions (all temperatures in degrees Celsius)

down by 40 per cent—a reduction in area almost equal to that of Brazil—and its winter thickness halved. The Arctic Ocean now absorbs 15 per cent more sunlight than it did just 30 years ago, as dark sea has swiftly replaced white ice. This reinforces warming across the region, which is expected to be ice free during the summer months as soon as 2030.

Arctic permafrost temperatures have climbed by 3°C on average, while ice on lakes and rivers at higher altitudes in the northern hemisphere now melts in spring two weeks earlier than a century ago. Northern hemisphere spring snow cover is already 8 per cent down on the 1922–70 mean and mountain glaciers continue to fade away, losing mass worldwide at a rate of up to 400 billion tonnes of ice every year. The loss of ice from the Greenland Ice Sheet, the world's second largest ice mass, has doubled over the last 20 years, and now averages a colossal 230 billion tonnes a year. The accelerating crumbling of this great mass of ice was highlighted in 2010 when a chunk of the Petermann Glacier, equal to a quarter the land area of Greater London, broke off and drifted off into the Arctic Ocean. At the opposite end of the planet, the loss of ice from Antarctica, too, has close to doubled over the last couple of decades; the rate of ice melting in summer is now ten times greater than 600 years ago and at its highest level for a thousand years.

Dramatically increasing the rate of melting of snow and ice means rising sea levels. Tide gauge data indicate that global sea levels rose by about 20cm during the 20th century: an average rise of 2mm a year. This rate has already started to accelerate, with annual rates of between 2.7 and 3.7mm a year recorded since 1993, a trend that is expected to accelerate through the course of this century and far beyond. According to the IPCC Fifth Assessment, even a worst-case scenario would only result in the ocean rising a metre or so by 2100, although a fair few in the climate science community feel that this could be a serious underestimate. Most of the recent and predicted rise comes from the thermal expansion of the oceans as they warm up or by the addition of water from the rapidly melting mountain glaciers. We still don't really have any true picture, however, of how rapidly the Greenland and Antarctic ice sheets will break up. If this happens catastrophically sea level could rise much more rapidly than predicted, with appalling consequences for coastal areas. If the entire Greenland Ice Sheet should melt, around 6.5m would be added to global sea level; the loss of the West Antarctic Ice Sheet (WAIS) adding a further 3.5m or so. The combined rise of 10m would see the homes of one-quarter of the current population of the United States swamped. At present, the WAIS appears to be stable, although severe warming over the next few centuries could result in its permanent disintegration and loss. The probability of the collapse and melting of the WAIS in the next 200 years has been put as high as 1 in 20. Should either the Greenland Ice Sheet or the WAIS melt fully, then virtually all the world's major coastal cities will find themselves under water.

Even without this, however, the effects of rising sea level in the next hundred years will be devastating for low-lying countries. A 1m rise, for example, would see the Maldives in the Indian Ocean under water, while a combination of rising sea level and ongoing land subsidence are forecast to result in a 1.8m rise in Bangladesh in just 50 years or so. This will see the loss of 16 per cent of the land surface, which supports 13 per cent of the population. Coastal flooding will also be enhanced by storm surges, with the number of people vulnerable across the world predicted to rise to two billion by mid-century. Because the oceans are so slow to respond to change, the problem of sea-level rise is not going to go away for a very long time. Even if we stabilized greenhouse gases in the atmosphere at current concentrations, sea level would continue to rise for a thousand years or more. It is a sobering thought that the last time global temperatures were as high as predicted for later this century—125,000 years ago at the peak of the interglacial before this one, known as the Eemian—sea levels were at least 4–6m higher than they are today.

It has become fashionable to blame every weather-related natural disaster on global warming. While it is difficult to determine if a particular weather event is a direct consequence of anthropogenic climate change, it can be done, and it has been convincingly shown that the severe UK floods in autumn 2000 and the extreme European (2003) and Russian (2010) heatwaves would otherwise have been extremely unlikely. There is now no doubt that extreme weather events of all types are increasing and rarely a week goes by without our television screens displaying harrowing images of the aftermath of flood, storm, wildfire, or drought (see Figure 8). Extreme precipitation events have increased significantly at high and mid-latitudes during the second half of the 20th century, and more rainstorms, floods, and windstorms are forecast. Current climatic characteristics are likely to be enhanced, so regions that are already wet will get wetter and those that are dry will suffer from prolonged and sustained drought. Northern Europe and the UK will therefore face more floods, while the North African deserts begin to creep towards southern Europe, and Australia begins to bake almost constantly beneath a blazing sun. Recent decades have seen an increase in the intensity of the most powerful Atlantic hurricanes. At the same time, tropical cyclones worldwide are getting stronger, with a 250 per cent increase observed in the number of storms with sustained wind speeds exceeding 175km an hour. Most climate models predict that, while it seems unlikely there will be a rise in the number of tropical cyclones as the world warms, the frequency of the most intense, and therefore potentially the most destructive, storms will increase. This is hardly surprising, as a rise in sea surface temperatures is the most likely primary driving mechanism for these more powerful storms, and this is climbing, on average, by about 0.1°C a year.

Where wind leads, waves often follow, and evidence is now coming to light of bigger and more powerful waves. Around the western and southern coasts of the UK, average wave heights— about 3 metres—have risen by over a metre compared to four

8. **Hurricane Sandy's storm surge breaches the coastal defences at New Jersey. Losses of US$65 billion made the storm the second costliest in US history, after Hurricane Katrina in 2005**

decades ago, while the height of the largest waves has increased by one third, to 10 metres. The unprecedented series of storms in the winter of 2012–13 spawned the biggest waves ever recorded in UK coastal waters. Although not yet attributed directly to global warming, the increased wave heights reflect changes in the weather patterns of the North Atlantic that in turn can be linked to the reorganization of our planet's weather system as it continues to warm. More coastal erosion is already taking its toll along many sections of the UK's most exposed coastlines: a situation that is likely to get much worse and that will undoubtedly be exacerbated by rising sea levels and storms.

There has been a great deal of discussion and debate in relation to how, if at all, climate change will influence El Niño, the second largest climate 'signal' after the seasons. An El Niño involves a weakening of the westward-blowing trade winds and the resulting migration of warm surface waters from the west to the eastern Pacific, devastating local fisheries and seriously disrupting the world's climate. The most recent research suggests that future warming is likely to double the frequency of severe El Niños, making a bleak picture even worse.

As the Earth continues to heat up, it looks as if it won't only be the seas and the skies that become increasingly agitated: the planet's crust will also join in; indeed, it is already doing so. In mountainous regions across the world, progressive warming is causing the permafrost that holds rock faces together to thaw, threatening villages, towns, and ski resorts with more frequent and more destructive landslides. The ice-covered peaks of Alaska used to host a major landslide about once every decade. During the last 15 years or so, however, the frequency has increased to one every three to five years. Half a world away, a gigantic, 100 million cubic metre slide of rock and ice, triggered in the Caucasus Mountains in 2002, obliterated farming communities and took more than 150 lives. Other major landslides at Mount Cook in New Zealand in 1997, at Monte Rosa in the Italian Alps in 2007, and at Mounts Stellar and Miller, Alaska, in 2008, all followed heatwaves: the frequency and duration of which climate change is projected to increase in future decades.

As the world continues to warm, there may be other, more fundamental responses from deeper levels within the Earth. Already, the reduced load acting on the crust as a consequence of the loss of ice mass in Alaska is freeing up the tectonic faults beneath, driving a rise in seismic activity. Some scientists are already wondering whether an increase in earthquakes in the North Atlantic region could be one legacy of the wholesale melting of the Greenland Ice Sheet. Looking back in time, particularly to the period following the last Ice Age, there is plenty of evidence for the solid Earth responding enthusiastically to rapid temperature rise, not only through earthquakes in the most unlikely places, but with an increase in volcanic activity, submarine landslides, and tsunamis.

Clearly, then, a major consequence of global warming will be a far more hazardous planet, few of whose inhabitants will escape scot-free. Already, things are getting rapidly worse, particularly along low-lying coasts and islands that are exposed to tropical cyclones and associated storm surges. According to Belgium's Centre for Research on the Epidemiology of Disasters, more than 244 million people were affected by natural disasters in 2011—mostly flood, storm, and drought—an astonishing 1 in 28 of the global population; and climate change has not really got steam up yet. Without doubt, all of us will be forced to embrace natural hazards as a normal, if unwelcome, part of our lives in the decades to come. Furthermore, the consequences of global warming stretch far beyond making the Earth more prone to natural catastrophes. Other dramatic and widespread changes are on the way that will have an equally drastic impact on all our lives. National economies will be knocked sideways and the fabric of our global society will begin to come apart at the seams, as agriculture, energy and water supplies, and human health and well-being become increasingly embattled.

A few countries will be able to adapt to some extent, but the speed of change is certain to be so rapid as to make this all but impossible for the most vulnerable nations in Asia, Africa, and elsewhere in the developing world. Against a background of soaring populations, falling incomes, and increasing pollution, there is no question that the impact of global warming will be terrible. One of the greatest problems will be a desperate shortage of water. Even today, 1.7 billion people—a third of the world's population—live in countries where supplies of fresh water are inadequate, and this figure will top five billion in just 25 years, triggering water conflicts across much of Asia and Africa. Alongside this, crop yields are forecast to fall in tropical, subtropical, and many mid-latitude regions, leading to the expansion of deserts, food shortages, and famine. The struggle for food and water will lead to economic migration on a gigantic scale, dwarfing anything seen today and bringing instability and conflict to many parts of the world.

In Europe and Asia trees come into leaf in spring a week earlier than just 20 years ago and autumn arrives 10 days later than it did. While this may seem beneficial, it will also encourage new pests to move into temperate zones from which they have previously been absent. Termites have already established a base in the southern UK, where, in places, temperatures are now high enough for malarial mosquitoes to survive and breed. In the tropics there will be an enormous rise in the number of people at risk from insect-borne diseases, especially malaria and dengue fever, while the paucity of drinkable water will ensure that cholera continues to make huge inroads into the numbers of young, old, and infirm. In urban areas, a combination of roasting summers and increased pollution will also begin to take their toll on health, particularly—once again—in poor communities where air-conditioning is out of the question. With land temperatures across all continents due to rise by up to 8°C by the end of the century, temperate and tropical forests, which currently help to absorb greenhouse gases, will start to die back, taking with them thousands of animal species unable to adapt to the new conditions. And not just the forests: grasslands, wetlands, coral reefs and atolls, mangrove swamps, and sensitive polar and alpine ecosystems will all struggle to survive and adapt, and many will fail to do so. Even our leisure activities will be affected. Not only will southern Europe become too dry for cereal crops, but it will also be too hot—in the summer months at least—for sun seekers. Prospects for the winter sports industry also look dismal, with most mountain glaciers likely to have dwindled to almost nothing by the end of the century, and snowfall much reduced.

Everything I have talked about so far is either already happening or has been predicted by powerful computer-driven climate models that are constantly being upgraded so as to forecast better what global warming holds in store for us. We must always be prepared, however, to expect the unexpected: drastic consequences that so far have been regarded as possible but not likely, or others that have simply not been thought of. One frightening possibility is that large sea-level changes due to global warming might trigger more volcanic eruptions, earthquakes, and giant landslides. Sounds crazy? As touched upon earlier, evidence from the past suggests that it might well be possible. When sea levels were rising rapidly following the end of the last Ice Age, the weight of the water on continental margins appears to have had a dramatic effect, causing volcanoes to erupt, active faults to move, and huge submarine landslides to collapse from continental shelf regions. While even the worst of worst-case scenarios does not envisage all the world's remaining ice melting due to human activities, which would result in a global sea-level rise in excess of 60m, the future rate at which sea levels climb could easily match the 7mm a year average of the post-glacial period.

The problem is that we don't know how big or how fast a rise is needed to see these phenomena happening again, although, interestingly, the Pavlov volcano in Alaska is induced to erupt in winter when low-pressure weather systems passing over raise sea level by just 15cm or so. There are other worries too. The accumulation of gases from the decomposition of organic detritus leads to the formation of what are called gas hydrates in marine sediments. These are methane solids that look rather like water ice, whose physical state is very sensitive to changes in their physical environment. A warming of just 1°C may cause rapid dissociation of the solid into a gaseous state, exerting increased pressure on the enclosing sediments and potentially leading to the destabilization and collapse of a huge sediment mass. This mechanism has been put forward as a trigger for the Storegga Slide: a gigantic submarine landslide off the coast of southern Norway, activated as the Earth continued to warm up 8,200 years ago. The collapse sent tsunamis up to 30m high surging across the Shetland Isles and onto the shores of north-east Scotland, leaving sandy deposits within the thick layers of boggy peat. As I address in more detail in my book *Waking the Giant*, it may well be that we are bequeathing to our children not only a hotter world, but also a more geologically dangerous one.

The good, the bad, and the downright mad

No one on the planet is going to escape the effects of anthropogenic climate change, and for billions the resulting environmental deterioration is going to make life considerably more difficult. It is too late now to put the clock back, but we can at least attempt to alleviate the worst impacts of warming. The question is, will we ever be able to achieve a worthwhile international consensus that allows us to do this with any degree of effectiveness? The Kyoto Protocol gave us some hope in 1997, with its goal of a 5.2 per cent reduction of greenhouse gas emissions (below 1990 levels) by 2008–12. Its success or failure can be easily measured by comparing global emissions in 1990 with those of 2012. Even a cursory glance at the figures—23 billion tonnes of carbon dioxide released into the atmosphere in 1990, compared to a record high of 31.6 billion tonnes in 2012—shows that the Protocol had a minimal impact on emissions growth. Many of the industrial nation signatories, particularly in Europe, may be able to pat themselves on the back for reducing their emissions over the period, but the fundamental point is that globally we are now pumping almost one-third more carbon dioxide into the atmosphere than 23 years ago. One of the reasons for this is that countries like the UK, whose home-grown emissions have gone down by 20 per cent over the past two decades, have accomplished this largely by 'exporting' much of them to countries such as China and India, who now make many of our consumer goods. This may make successive UK governments feel good, but it is irrelevant in the context of the broader picture. The Earth's climate does not recognize which country the carbon dioxide comes from and in fact, over the period, the UK's carbon footprint has grown by 10 per cent.

There is much talk at the moment about how fracked gas is helping to reduce emissions in the United States, by taking the place of coal. This might be true, the country posting, in 2012, its lowest emissions since the mid-1990s. Unfortunately, the coal that is not

41

being burnt in the USA is now free for export so that it can be burnt elsewhere, so there is absolutely no benefit at the global level. Furthermore, fracking itself releases significant quantities of the potent greenhouse gas methane. Thus, the worldwide adoption of fracked gas as an energy source would be a catastrophe in terms of keeping climate change under control, locking countries, as it would, into carbon-based energy production for the foreseeable future instead of embracing renewable, low-carbon, energy sources like the sun, wind, and tides.

The more global warming continues to grab the limelight, so the more we hear from what I shall call the 'techno-fix tendency'. These are the scientists, industrialists, and others who are keen to try and 'solve' the climate change problem by engineering—or 'hacking'—the planet. Some of the earliest 'geoengineering' schemes for mitigating warming were wild and wacky, such as placing giant reflectors in space to divert solar radiation or, even more fantastically—and heaven forbid—diverting a comet or two past the Earth, utilizing their gravity to swing the planet out into an orbit further from the Sun. Today, however, geoengineering is becoming 'respectable', at least in some circles, and options are under serious consideration that we already have the technology to make possible.

Top of the list of geoengineering solutions is the idea that we mimic a large volcanic eruption by pumping large quantities of sulphur aerosols into the stratosphere. As is known from the cooling that follows major explosive eruptions, such as that of Tambora in 1815 and, more recently, Pinatubo in 1991, sulphur aerosols are very effective at absorbing and reflecting incoming solar radiation, thereby cooling the lower atmosphere and the surface. Spraying aerosols into the stratosphere from giant tethered balloons is now being seriously considered as a relatively cheap means of keeping rising temperatures under control that has a reasonable possibility of success. The problem is that computer models have shown that interfering with the climate in this way can have drastically different effects at the regional scale, especially with regard to rainfall; so while some countries might be inundated by flood waters, others would be visited by famine and drought as their normal rainfall patterns are disrupted. In addition, while such a scheme might temporarily slow warming, it would not curtail other aspects of climate change, in particular the increasing acidification of the oceans, now more acid than for 300 million years or more.

Whatever they might say in public, I have no doubt that the reason many in business and government are coming to embrace geoengineering 'solutions' to climate change is because they think that free market, consumer-led, growth can continue without the inconvenience of having to think about cutting back on the greenhouse gas emissions that result. As a means to provide a breathing space while drastic cuts are made in emissions, geoengineering has something to be said for it. As an alternative to emissions cuts it would be a disastrous way forward. Unless greenhouse gas emissions are cut simultaneously, we would be digging ourselves ever deeper into an enormous hole that would see us having to continue to pump out more and more sulphur gases in order to counteract ever-increasing levels of carbon dioxide and other warming gases in the atmosphere. We would simply not be able to stop without the planet experiencing catastrophic warming almost overnight as the full force of accumulated greenhouse gases hit home with a vengeance.

In reality, geoengineering schemes are not the answer. If we are to have any impact on global warming we have to change our lifestyles at the global level, moving away from a disposable society and towards one that promotes and rewards the most effective and efficient use of available energy and resources. Tackling global warming is inextricably linked with the widespread adoption of sustainable development. Climate change will bring to an end the world as we have known it through dramatic modifications to our environment, but if the situation is not to continue to slide it must

43

also provide the incentive and impetus for changing the way we live. In the developed world we have no choice but to cut fuel consumption, invest in renewable energy sources, recycle on an immensely greater scale, and produce locally as much as possible rather than shipping fruit and vegetables halfway around the planet. Much as I can understand their resistance, governments of developing countries cannot and must not follow the wasteful route to industrialization that Europe and North America have taken, for the simple and logical reason that if they do, they—and their people—will be the ones who suffer most.

As I write these words, the launch of the IPCC's Fifth Assessment Report on the physical basis of climate change is one of the news items of the day. Not only, however, is it not the lead story, but the reports fail to get over to the viewers and listeners any sense of the desperate urgency of the situation. The journalists and news readers could equally well be talking about the latest royal birth or the sports results for all the significance attached to the pieces. Notwithstanding the expected denialist spoiling from much of the right-wing media, who still insist that climate change is either not happening or is not our fault, it is extraordinary, and dispiriting, that the thrust of most of the news stories seems to be that the latest IPCC report concludes that climate change is happening and that it is 'almost' certain to be the result of human activities. Haven't we been here before? If, after nearly a quarter of a century, this is as far as we have travelled in terms of getting the message across, then maybe there is, after all, no hope. Certainly, prospects for the development of a consensus that recognizes the urgency of tackling climate change seem increasingly dire. Deniers have jumped upon the fact, while conveniently ignoring the observation that each decade is hotter than the last, that the rate of warming seems to have slowed somewhat over the last decade and a half. The science suggests that this is likely to be because the oceans are taking up more of the heat, but since when have these people ever been interested in the science? Slowly, the deniers' intentional obfuscation and muddying of the waters are confusing

44

the true picture, with the result that, according to surveys, fewer people now think that climate change is real. Even those who do have an increasing tendency to think, like Mr Micawber, that something will turn up to make everything right.

Slowly, slowly, things are changing. Sustainable and low-carbon energy sourced from the sun, wind, water, and biomass are gradually making greater contributions to the energy mix, while electric vehicle production and ownership is creeping up, but it is all too little, too late. To have even a 50:50 chance of keeping global average temperatures below the 2°C guardrail, widely recognized as the threshold above which climate change becomes all-pervasive, catastrophic, and—quite possibly—run-away, the latest IPCC report says we can't emit more than 820–1,445 tonnes of carbon dioxide and other greenhouse gases. At current emissions rates, we would use up this ration in somewhere between 16 and 29 years. Already, the Chinese have indicated that it would be impossible for them to reduce their reliance on fossil fuels in line with what is needed, and other industrialized and industrializing nations are making similar noises. In short, it isn't going to happen. Global emissions will roll through the targets that give us an evens chance of preventing dangerous climate change, and will keep on going; no one knows for how long or, ultimately, to what levels. All this despite recent research by lauded climate scientist James Hansen and others, which suggests that if we burn all fossil fuels, global average temperatures will ultimately climb by 16°C, leaving most of our world a roasting hell-hole too hot for humans to inhabit.

It may well be, then, that the only way urgent action will ever be undertaken to seriously tackle climate change is if society is shaken out of its complacency by some Earth-shattering climate catastrophe that no one can ignore. One threat clearly stands out from the rest: a sudden major 'burp' of methane from the thawing Arctic permafrost. More than a trillion tonnes of methane is now trapped beneath the surface of the East Siberian Arctic

45

Shelf, ready and waiting to explode into the atmosphere. Within a decade, the near instantaneous release of 50 billion tonnes of methane is thought to be possible, increasing 12-fold the methane content of the atmosphere, virtually overnight. The consequences would be immediate and catastrophic, global average temperatures climbing by around 1.3°C. It would be as if global warming had been instantly brought forward by 35 years, so that the critical 2°C dangerous climate change guardrail was bypassed at a stroke. To say nothing of the impact of such a methane outburst on the climate and the environment—severe drought, extreme heatwaves, intense storms, rapid sea-level rise, crop losses, and the rest—one forecast predicts that such a climate cataclysm would cost the global community something like US$60 trillion, not far short of a year's GDP. The potential worldwide impact has been described as an economic 'time bomb' and, according to Arctic methane researcher Natalia Shakhova, at the University of Alaska, one that could go off 'at any time'. It is perfectly possible that when and if the big burp ever happens— and scarily, plumes of methane a kilometre across are already bubbling to the surface—it might finally trigger some serious action to tackle to climate change. By this time, however, it will probably be too late.

Chapter 3
The Ice Age cometh

The furnace or the freezer?

One of the main reasons for a growing disillusionment with science amongst the general public is the perception that scientists are always arguing with one another and constantly changing their minds. It is no use explaining that this is how science progresses: through battles between competing theses until the accumulation of evidence ensures that one triumphs and becomes an accepted paradigm. People want scientists to agree; to present a united front; and to tell them what is true and what is not. They want this because it makes life that much easier and gives them that much less to worry about. If you are concerned about your career or your marriage you don't want to think about whether GM crops are good or bad, or whether it is safe to have your kids immunized, or whether your children's children are going to fry or freeze. Once again, however, the scientific consensus at least appears to have done another U-turn over the last couple of decades. As we saw in the last chapter, the fact that the Earth is warming up rapidly is now unequivocal and our polluting activities are clearly to blame. As recently as the 1980s, however, the big question in climatological circles—at least as portrayed in the populist media—was when can we expect the next Ice Age? So what has changed? Well, actually, not much. Few climate scientists of repute in the 1980s gave any credence at all to the idea that a return to Ice Age conditions was imminent, and concern was already building about how human actions were heating the planet. What has changed is the recognition that anthropogenic warming and its associated climatic impact may have a role to play at a critical time of natural transition when our interglacial world is due—within a few thousand years or so—to give itself over once again to advancing ice sheets. Problematically, though, researchers are not quite sure what this role will be, and although, intuitively, you might expect global warming to delay or even fend off entirely the next Ice Age, some scientists have suggested that the ongoing dramatic rise in temperatures may bring colder conditions to parts of our world, at least for a time. Even if the latter is shown not to be the case—as is increasingly likely—we still have a problem. Knowing that we are coming towards the end of an interglacial period and that a new age of ice is on the way, should we not be trying actively to keep our planet warm? Should we not welcome global warming with open arms? In other words, we are currently faced with a stark choice that is only rarely voiced during the great global warming debate. How do we wish our familiar, climatologically comfortable world to end—in the furnace or in the freezer?

How to freeze a planet

During the Earth's early history the surface boiled with lava oceans and exploding volcanoes, and although temperatures fell dramatically as prevailing geological processes moderated, our planet has been bathed in warmth for most of its 4.6 billion year history. Occasionally, however, a fortuitous combination of circumstances has heralded the formation of enormous ice sheets that have transformed a balmy paradise into a freezing hell. Artists' impressions and television documentaries have ensured that most of us are familiar with the last great Ice Age, when mammoths roamed the tundra and our pelt-covered ancestors struggled to eke out an existence from a frozen world. Only relatively recently, however, have studies of ice-related rock formations around the world brought to light a far more ancient and much more terrible period of refrigeration; a time when our planet was little more than a frozen snowball hurtling through space. Long, long ago, during a geological episode appropriately referred to as the Cryogenian (after *cryogen* for freezing mixture), the Earth found itself at a critical threshold in its history. It had cooled substantially since its formation over three and a half billion years earlier and now the problem was keeping itself warm. At this time, between about 850 and 635 million years ago, the Sun was weaker and the Earth was bathed in some 6 per cent less solar radiation than it is now. Furthermore, the concentrations of greenhouse gases that are now heating up our planet—primarily carbon dioxide and methane—were not sufficiently high to hold back entirely the bitter cold of space. Huge ice sheets rapidly formed and pushed towards the equator from both poles, encasing all or most (this remains a bone of contention) of the Earth in a carapace of ice a kilometre thick. As the blinding white shell reflected solar radiation back into space, temperatures fell to −50°C and prospects for an eternity of ice seemed strong. But something must have happened to break the ice, as it were, otherwise I would not be here today to tell you about it. In fact it seems that these 'snowball' conditions may have developed up to half a dozen times, succumbing, on each occasion, to a return of warmer climes.

Just how the Earth managed to escape the clutches of the ice no one is quite certain, but it looks as if volcanoes might have been the saviours. After millions or even tens of millions of years of bitter cold, the enormous volumes of carbon dioxide pumped out by erupting volcanoes seem to have generated a sufficiently large greenhouse effect to warm the atmosphere and melt the ice. Extraordinarily, life came through this particularly traumatic period of Earth history bruised and battered but raring to go, and hard on the heels of Snowball Earth's final fling came the great explosion of biodiversity that marked the start of the Cambrian period 541 million years ago. Compared to the great freezes of the Cryogenian our most recent Quaternary ice ages come across as

49

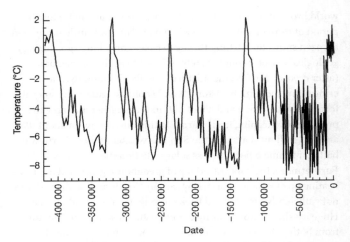

9. **Temperature changes over the past 420,000 years show that the Earth has been much colder than it is now for most of the time**

rather small beer. Nevertheless, although they affected smaller areas of the Earth's surface, these latest bouts of cold (see Figure 9) were crucial because they coincided with the appearance and evolution of our distant ancestors. Furthermore, they may yet have a role to play in the future of our race.

During more recent Earth history the Sun's output has been significantly higher than during the Cryogenian and the level of carbon dioxide and other greenhouse gases has also been higher. Why then, at the end of the Miocene period about 10 million years ago, did glaciers once again begin to form and advance across much of the northern hemisphere? And more importantly, why, around three million years ago, did the southward march of the ice intensify? This remains a particularly hot topic in the fields of Quaternary science and environmental change and a detailed analysis of competing theories is beyond the scope of this book. Suffice it to say that explanations for the twenty or more ice ages

50

that have gripped the Earth during the last two million years include disruption of atmospheric circulation and rainfall patterns due to uplift of the great Himalayan mountain belt, and the drastic modification of the global system of ocean currents by the emergence of the Panama Isthmus.

Although one or both of these spectacular geophysical events may have contributed to a picture of increasing cold, the ice was already on the move, and we need to look elsewhere for the true underlying cause. What, in other words, turns ice ages on, and—just as importantly—what turns them off? This problem has intrigued scientists for many years and the solution was first put forward by the Scottish geologist James Croll as long ago as 1864 and expanded upon by the Serbian scientist Milutin Milankovitch in the 1930s. The Croll–Milankovitch Astronomical Theory of the Ice Ages proposes that long-term variations in the geometry of the Earth's orbit and rotation are the fundamental causes of the blooming and dying of the Quaternary ice ages (see Figure 10). In order for an ice age to get going, the astronomical theory requires that summers at high latitudes in the northern hemisphere are sufficiently cool to allow the preservation of winter snows. As more and more snow and ice accumulate year on year, so the reflectivity or albedo of the surface is increased, causing summer sunshine to have even less impact and accelerating the growth of ice sheets and glaciers. But how are the northern hemisphere summers cooled down in the first place? This is where the astronomy comes in. Cooler summers at high latitudes result from a reduction in the amount of solar radiation falling on the surface, and this in turn depends upon both changes in the tilt of the Earth's axis *and* variations in its orbit about the Sun.

If the Earth's axis was not tilted then we would not experience the seasons. During the northern hemisphere summer, for example, the North Pole is tilted towards the Sun, allowing more direct solar radiation to reach the surface in the northern hemisphere and raising the temperatures. In contrast, during the winter, the North Pole is tilted away from the Sun and the long, balmy days of summer are replaced by the cold and dark of a northern hemisphere winter. Now the southern hemisphere receives more direct sunlight with the result that those down under bask in warmth while the north shivers beneath gloomy skies. Although the tilt of the Earth's axis averages about 23.5 degrees, it is not constant. Like a spinning top, the Earth wobbles—or precesses—about its axis of rotation over a period of between 23,000 and 26,000 years. Furthermore, this wobble causes the amount of tilt to vary between 22 and 25 degrees over a period of 41,000 years. At times of least tilt, winters are actually milder, but more importantly, high latitudes receive less direct solar radiation and become cooler, making the survival of winter snows and the growth of ice sheets easier. On top of this there is another so-called astronomical forcing mechanism that contributes to the onset of ice age conditions. Like all planetary bodies, the Earth follows an elliptical rather than a circular path around the Sun, whose shape varies according to cycles of between 100,000 and 400,000 years. At the moment the Earth's closest approach to the Sun occurs in January, when the North Pole is pointing away from the Sun, resulting in slightly colder northern hemisphere winters. Just 11,000 years ago, however, this closest approach—or perihelion—occurred in July, giving a small temperature boost to northern hemisphere summers.

Before this gets too complicated let me try and draw things together. Regular and predictable cycles—known as Milankovitch Cycles—are recognized in the behaviour of the Earth's tilt and its orbit over periods of thousands to hundreds of thousands of years, and these cycles control the amount of solar radiation reaching the Earth's surface and therefore its temperature. At times, a number of cycles may coincide so as to depress summer temperatures at high latitudes to a degree sufficient to allow the accumulation of winter snows. On its own this could not result in the huge ice sheets that have dominated the northern hemisphere for much of the last few million years, but as the area covered by snow and ice grows, so more and more sunlight is reflected back into space, accelerating the cooling process. This—in essence—is how ice ages start. Conversely, at other times, the various cycles cancel one another out, the planet warms as a result, and the ice sheets retreat to their polar fastnesses.

Although Milankovitch and later researchers who have addressed the issue have been able to explain the mechanics of the ice ages and their periodicity, they have been less successful in deciding why these icy episodes appeared on the scene around ten million years ago, rather than being apparent throughout Earth history. An answer to this may lie, however, in the carbon dioxide level of the Earth's atmosphere, which has been steadily falling over the last 300 million years, from about 1,600 parts per million to just 279 ppm prior to the industrial revolution. It has been suggested that perhaps only when the level of carbon dioxide in the atmosphere drops below a critical threshold level—say of 400 ppm—is astronomical forcing sufficient to initiate the cycle of warm and cold that characterizes the ice ages. This begs the question that with carbon dioxide levels in the atmosphere exceeding this value in 2013 and still heading remorselessly upwards, have we seen off the ice ages for ever? I shall return to this later.

In the meantime, on the basis that there is at least a fair chance that we will have to face them again at some time in the future, let's examine what conditions were like in the depths of the last Ice Age. As temperatures started to fall around 120,000 years ago, so more and more of the planet's water found itself locked up in mountain glaciers, polar sea ice, and expanding continental ice sheets in the northern hemisphere, with the result that sea level began to fall dramatically. Ice swept southwards towards the equator on at least four occasions over this period, with the peak of ice cover being reached a mere 15,000–20,000 years ago.

At this time sea level was a good 120m or so below what it is now—about the height of a 40-storey building—exposing new

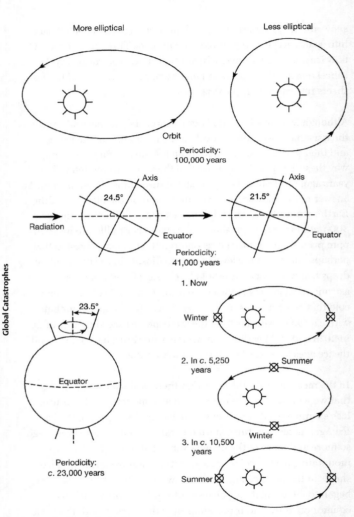

10. **Milankovitch Cycles control the timing of the ice ages; variations in the shape of the Earth's orbit around the Sun (top), changes in the tilt of the Earth's axis (middle), and precession about the Earth's axis (bottom)**

land bridges between continents and facilitating the migration of both animals and our distant ancestors. One of these land bridges developed across the Bering Straits, allowing people from Asia to cross into North America, from which, eventually, they colonized the whole of the New World. Just 600 generations ago, then, the north of our planet was in the steely grip of full glaciation with a third of all land covered by ice and 5 per cent of the world's oceans frozen. Compared to the 21st century, the environment at the height of the last Ice Age was desperately hostile, with mean global temperatures down 4°C on today but far lower at high latitudes in the north. In the UK, temperatures were reduced by between 15°C and 20°C, transforming the country into a frozen wasteland with great sheets of ice reaching as far south as the River Thames and beyond. Some of the most inhospitable conditions were, however, to be found in North America, where temperatures across huge areas were 25°C lower than today and ice fields kilometres thick ensured that life was largely impossible. Remarkably, just as it seemed as if the world might be returning to the 'snowball' state of the Cryogenian, a surprising change took place. The planet started to warm rapidly, melting the great ice sheets far more quickly than they took to form. Meltwater poured into gigantic lakes at the margins of the ice fields, which, in turn, emptied into the oceans, raising sea level and inundating land exposed just a few thousand years earlier.

By 12,000 years ago, sea level was rising far more rapidly than even the most pessimistic forecasts for the next century, possibly by as much as 10m or so in a couple of centuries, and all the time the climate was becoming warmer and warmer—well, almost all the time, that is. The journey from the depths of Ice Age to the current balmy interglacial was a rather bumpy one, and on more than one occasion the ice made a concerted attempt to reclaim centre stage. A little under 13,000 years ago, for example, the rapid retreat of the ice was stopped in its tracks as a new blast of cold initiated a 1300-year-long freeze, known as the Younger Dryas (to distinguish it from an earlier and less severe cold phase called the Older Dryas). No one is certain what caused this sudden cold snap but one suggestion is that the culprit was a colossal discharge of fresh water from long-gone Lake Agassiz, one of the gigantic glacial meltwater lakes that had accumulated in North America. The catastrophic emptying of this lake into the St Lawrence River, and thence into the North Atlantic, may have disrupted currents carrying warmer waters into polar regions, allowing the climate at higher latitudes to cool and ice to form once again. The Younger Dryas and similar post-Ice Age cold snaps teach us a couple of important lessons that we would do well to remember as our own world undergoes dramatic climate change. First, the switch from warm to cold and vice versa can occur extraordinarily rapidly, perhaps—according to the latest research—in just one or two years. Second, the disruption of ocean currents can have serious and far-reaching consequences for climate change. Some worrying implications of the latter I shall address in more detail later in this chapter.

Charles Dickens, White Christmases, and the Little Ice Age

It seems likely that almost everyone who has read this far is familiar with *the* Ice Age, but what about the *Little* Ice Age? This is the term used by climatologists to describe a cold period that lasted from at least 1450—and possibly 1200—until between 1850 and the start of the 20th century. During this time, glaciers advanced rapidly, engulfing alpine villages, and sea ice in the North Atlantic severely disrupted the fishing industries of Iceland and Scandinavia. Eskimos are alleged to have paddled as far south as Scotland, while the once thriving Viking community in Greenland was cut off and never heard from again. Annual mean temperatures in England during the late 17th century were almost 1°C lower than for the period 1920–60, leading to bitter, icy winters in which 'frost fair' carnivals were held regularly on a frozen River Thames and snowfall was common (see Figure 11). The snowy winters described in many of the works of Charles Dickens

56

may well be a reflection of this colder climatic phase, and they have certainly done much to nurture our constant expectation of—and yearning for—an old-fashioned 'white Christmas'.

Just what was the cause of the Little Ice Age remains a matter of intense debate. Clearly, however, as most of the cold snap occurred prior to the industrial revolution there can be no question of human activities having played a role. Despite this, it is vital that we understand the Little Ice Age in the context of global warming, if for no other reason than if we don't appreciate the natural variations of our planet's recent climate it is well nigh impossible to unravel the effects arising from human activities. In fact, the Little Ice Age was not the only significant departure from the climatic norm—if there is such a thing—in historical times. Immediately prior to this cold snap, Europe, at least, was revelling in the so-called Medieval Warm Period. This time of climate amelioration, between about AD 1000 and 1300, saw grapes grown in the north of England—as they are again in today's warming climate—while the Norse settlers of Greenland were able to graze their livestock in areas later buried beneath ice. The emergence of the world from the Little Ice Age towards the end of the last century, coincident with the acceleration of industrialization on a global scale, has contributed in no small way to current arguments on the causes of contemporary warming.

As I noted in the last chapter, the overwhelming scientific consensus views global warming as being anthropogenic in nature, but those sceptics and deniers who—for one reason or another—will only accept an entirely natural explanation try to explain the current warming in terms of the planet coming out of the Little Ice Age and entering another warm episode analogous to the Medieval Warm Period. Although the available evidence irresistibly supports a human cause rather than a purely natural driver, there can be no doubt that the impact of human activities is superimposed upon a natural variability that in the recent past has resulted in significant climate change. But

57

what is the cause? One of the most likely culprits is the Sun, whose output continues to vary on time scales ranging from 100 to 10,000 years. For example, the two coldest phases within the Little Ice Age corresponded closely with two episodes of reduced solar activity: the Spörer Minimum between AD 1400 and 1510 and the Maunder Minimum from 1645 to 1715. During these times, observation records reveal that virtually no sunspots were visible and auroras were almost non-existent, suggesting a fall in the rate of bombardment of the Earth by radiation from the Sun. While solar physicists estimate that the Sun during the Maunder Minimum may have been just a quarter of one per cent dimmer than it is today, this might have been sufficient to cause the observed cooling. In addition, other factors may also have made a contribution, and a recent theory has given elevated levels of explosive volcanic activity at the time—including the great 1815 eruption of Indonesia's Tambora volcano—at least a supporting role in the Little Ice Age cooling. As I shall discuss further in the next chapter, large volcanic explosions are particularly effective at injecting substantial volumes of sulphur dioxide and other sulphur gases into the stratosphere—that part of the atmosphere above 10km or so. Here they mix with atmospheric water vapour to form a fine mist of sulphuric acid aerosols that cuts out a proportion of incoming solar radiation and leads to a cooling of the troposphere (the lower atmosphere) and surface.

A very British ice age

The more we learn about past climate change, the more it becomes apparent that dramatic variations can occur with extraordinary rapidity. The switch—seemingly within just a year or two—from increasingly clement conditions to the bitter cold of the Younger Dryas, 12,800 years ago, demonstrates this, as does the similarly rapid transition from the Medieval Warm Period to the Little Ice Age. Equally disturbing is the tendency for the climate to flip suddenly from one extreme to another when it is under particular stress, as it is at the moment from anthropogenic

11. During the Little Ice Age winters were often cold enough for ice fairs on the River Thames

warming. This once again raises the question I posed at the beginning of this chapter—is there any way that current global warming can actually bring a return to colder conditions? While this would seem to be counter-intuitive, it is something to which climate scientists have given serious consideration, most notably in relation to the UK and north-west Europe. The only reason tropical palms are able to thrive in western Ireland and south-west England is because the Gulf Stream carries northwards warm water from the Caribbean. As a result, the UK and Ireland are substantially warmer than comparable latitudes in eastern Canada, which have to put up with sub-Arctic conditions. But what would happen if the supply of warm water from the south were shut down? It is highly likely that the British climate— and perhaps that of much of north-west Europe—would become bitterly cold, and some have suggested it could even rival that of Svalbard (formerly Spitsbergen), the ice-shrouded islands off east Greenland where the polar bear is king.

One of the ways of weakening or shutting down the Gulf Stream is by short-circuiting it through releasing huge quantities of cold fresh water into the North Atlantic, and this is just what has been

59

predicted by a number of different climate models developed to look at the impact of global warming in this century and beyond. Some forecasts have suggested that a 2–3°C global average temperature rise, which is virtually certain by 2100 if not well before, will result in a close to 1 in 2 chance that the Gulf Stream will shut down or slow dramatically. In little more than half a century, then, is it possible that the seas around the UK could be significantly cooler, altering prevailing weather patterns and bringing colder conditions to the region? While the rest of the world roasts, could the North Atlantic region conceivably start to slide into a freeze very much more bitter than the Little Ice Age? Is it even possible that this might just be the start, the knock-on effects of changes to the ocean circulation in the North Atlantic spreading to overwhelm the current warming and bringing a return of the ice across the northern hemisphere? In short, could our warming activities actually hasten, rather than postpone, the arrival of the next Ice Age?

Out of the frying pan into the fridge?

In terms of the Milankovitch Cycles, our planet is already primed for the end of the current interglacial and a return to full ice age conditions. Some believe that all that is needed is a trigger: a sudden shock to the system that will knock the climate out of equilibrium and set it wobbling, before it collapses into an altogether less friendly state. It is highly questionable whether global warming can provide a shock of the appropriate magnitude, but research over the last decade or so has at least shed light on how warming can indeed lead to a cooler future, at least for some. Once again, the key seems to lie in the ocean circulation system of the North Atlantic, which appears to be closely bound up with past switches from warm to cold episodes and vice versa. The Gulf Stream that most people are familiar with is actually only one part of a system of currents known by a variety of names, of which the Atlantic Meridional Overturning Circulation (AMOC) is probably the most revealing. As the warm, salty waters of the Gulf Stream head northwards they cool and consequently become more dense. As a result, by the time they have reached the Arctic Ocean they have sunk to form a cold, deep-ocean current that heads south once more to join the wider system of ocean currents known as the Thermohaline Circulation or, more simply, as the *Global Conveyor*.

It now looks as if the workings of the AMOC are seriously disrupted whenever cold conditions grip the northern hemisphere. During the Younger Dryas, for example, the circulation appears to have been severely reduced, lowering north European average temperatures by as much as 10°C. Recent evidence on ocean temperatures and salinities, gleaned from studies of the shells of tiny marine organisms known as foraminifera, also points to a much weaker Gulf Stream at the height of the last Ice Age some 20,000 years ago. Then, it seems, the Gulf Stream had only two-thirds of its current strength, suggesting that the entire circulation system was comparably weakened. The question is, did this weakening have a role to play in the triggering of the last Ice Age, or was it merely a consequence? No one really knows, but there is a general feeling that a weakening of the circulation results in much colder conditions in the northern hemisphere and that such a weakening appears to be associated with large influxes of cold water into the North Atlantic. Due to the extraordinarily rapid melting of Arctic sea ice and the Greenland Ice Sheet, this is just what we are seeing today.

During the Younger Dryas, 12,800 years ago, the release of huge quantities of water from glacial lakes resulted only in a short-lived cold snap of 1,300 years or so. Then, however, the Earth was at a point in the pattern of Milankovitch Cycles when temperatures were on the way up. Now, we are poised at a natural transition between the present interglacial and the next Ice Age, and without the polluting effects of human activities temperatures might reasonably be expected to be on the way down (see Figure 12). Is it not reasonable to at least consider, then, that the current deluge of cold, fresh water into the Arctic Ocean may eventually trigger not just

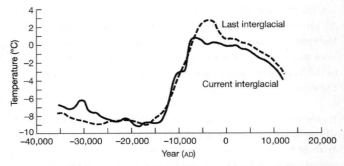

12. Comparisons of temperatures in this interglacial period and the last suggest that we are already well on the way to the next Ice Age

a brief period of cold in north-west Europe, but a new Ice Age affecting the entire northern hemisphere? So far, the good news is that—as yet—there is no sign of the wholesale melting of Arctic ice causing the Gulf Stream and its associated currents to slow. There was a flurry of concern in 2005, when a paper was published suggesting that a sudden 30 per cent slowdown may have occurred. Since then, however, more detailed observations have indicated that the AMOC is prone to speeding up and slowing down over the course of just a few years, with no overall pattern yet obvious. Notwithstanding this, it has been proposed recently that the rapid melting of Arctic sea ice might have something to do with the recent run of cold UK winters, through playing havoc with the position of the high-altitude band of super-fast winds known as the Jet Stream. In its 2013 Fifth Assessment Report, the IPCC forecasts that while the AMOC may slow by as much as 50 per cent this century, it is 'very unlikely' that it will suddenly grind to a halt. Furthermore, a UK Met Office simulation reveals that, even should the AMOC shut off suddenly, the resulting bubble of cold would be pretty much confined to the UK and Scandinavia. Elsewhere, including across the Channel in mainland Europe, temperatures would just keep on climbing.

It seems, therefore, that the return of the great ice sheets is not imminent and that it is the furnace not the freezer that we should be most concerned about as we look ahead to the remainder of this century and beyond. Notwithstanding this, however, we can see how bad a time this is for us to be experimenting with the global climate by comparing the temperature profile of this interglacial with that of the last. It is rather sobering to see that the natural temperature trend is already downwards, and that this fall has already been going on for several thousand years. A recent study by scientists from the UK, the United States, and Norway suggests that, in the absence of human-induced global warming, our interglacial would come to an end sometime in the next 1,500 years, but this is being staved off by higher and climbing greenhouse gas levels in the atmosphere. What will happen, though, if the world comes to its communal senses and the amount of carbon dioxide and other greenhouse gases that we pump into the atmosphere is slashed? Well, you have seen the graphs; eventually, the ice will be back with a vengeance. It is simply a matter of whether we prefer to take the icy plunge sooner, or spend a while steaming in the sauna first. Whichever choice we make, there is no denying that life for our descendants will become increasingly hard, when and if the ice returns. Life in Europe, North America, Russia, and central and eastern Asia would be pretty much impossible, prompting mass migrations southwards accompanied, undoubtedly, by bloody wars fought over living space and resources. The climate of Ice Age Earth is simply not suited to sustaining a population of 8–10 billion, or more, and widespread famine alongside civil strife is certain to lead to a severe culling of the human species. There is no question that our race will survive, as it did the last time that the ice left its polar fastnesses, but it is likely to be but a pale shadow of its former self.

Chapter 4

The enemy within: super-eruptions, giant tsunamis, and the coming great quake

Hell on Earth

The staggering power of even a common-or-garden volcano in full eruption is difficult to comprehend and impossible to protect against: evidence the cataclysmic detonation that blew Krakatoa (Indonesia) apart in 1883, killing around 36,000 of the inhabitants of Java and Sumatra, or the titanic blast that tore the Greek island of Thera to pieces one and a half millennia before the birth of Christ. Imagine, then, the capacity for mayhem of one of Nature's greatest killers, the volcanic super-eruption: a blast more than a hundred times bigger than Krakatoa.

Fortunately for us, super-eruptions are far from common, and it is estimated that throughout the last two million years of Earth history there have been, on average, just a couple of such blasts every hundred millennia, the last shattering the crust at Taupo in New Zealand's North Island 26,500 years ago. This does not, however, mean that we can sit back and relax for another 24 millennia or so. Like buses, natural phenomena display scant regard for a timetable, so the next super-eruption could be 100,000 years away—or it could burst upon us just a few years down the line. The really scary thing is that, unlike 'normal' volcanic blasts, there is no possibility of avoiding the devastating consequences of a super-eruption. Those of us tucked away in the most geologically friendly countries will still find our cosy world turned upside down by the next one, even if it occurs on the other side of the planet.

This is a legacy of the colossal scale of super-eruptions, which score an eight on the so-called Volcanic Explosivity Index or VEI, developed to measure and compare the sizes of volcanic blasts. Because (like the better-known Richter Scale for earthquakes) the VEI is logarithmic, each point on the scale represents an eruption ten times larger than the one immediately below. Thus a VEI8 is ten times bigger than a VEI7 and 100 times greater than a VEI6 event, such as Krakatoa or the 1991 eruption of Mount Pinatubo in the Philippines. Even the smallest of super-eruptions would be 1000 times bigger than the 1980 VEI5 blast of Mount St Helens (Washington State, USA), which itself was big enough to cause regional disruption. For the nearest thing to a super-eruption in modern times we have to go back almost two centuries to 1815— the year of the battle of Waterloo. As the armies of Wellington and Napoleon jockeyed for position across Europe, on the distant Indonesian island of Sumbawa, the long-dormant Tambora volcano ripped itself apart in a gargantuan VEI7 blast that may have been the largest since the end of the Ice Age. Sir Stamford Raffles, then British Lieutenant Governor of Java, reported a series of titanic detonations loud enough to be heard in Sumatra 1,600km away. When the eruption ended, after 34 days, it left 12,000 dead. In the ensuing months, however, a further 80,000 Indonesians succumbed to famine and disease as they struggled to find food and uncontaminated water across the ash-ravaged landscape.

Utterly devastating though the Tambora event no doubt was to the people of Indonesia, its immediate effects were nonetheless confined to one part of South East Asia. Indirectly, however, much of the world was to suffer the consequences of this huge blast. Along with some 50 cubic km of ash, the climactic explosions of the Tambora eruption also lofted around 200 million tonnes of sulphur-rich gases

65

into the stratosphere, within which high-altitude winds swiftly distributed them across the planet. The gases combined readily with water in the atmosphere to form 150 million tonnes of sulphuric acid aerosols—tiny particles of liquid that are very effective at blocking out solar radiation. Within months the northern hemisphere climate began to deteriorate and temperatures fell to such a degree that 1816 became known as the 'year without a summer'. Global temperatures are estimated to have fallen by around 1°C or so, causing summer frosts, snows, and torrential rains. The miserable weather conditions seem to have set just the right mood for Mary Shelley's vivid imagination to give birth to its most famous offspring, *Frankenstein*, while the spectacular ash- and gas-laden sunsets are said to have inspired some of J. M. W. Turner's most brilliant works.

Certainly the weather conditions in Europe and North America during 1816 were awful, but could a volcanic eruption in a far-off part of the world really change the climate so much as to cause a breakdown in society and end the world as we know it? Evidence from the past suggests that it can. Far back in the geological record—during the Ordovician period some 450 million years ago—an enormous volcanic explosion in what is now North America ejected sufficient ash and pyroclastic flows to obliterate everything across an area of at least a million square km. This is broadly the size of Egypt or four times the area of the UK. In addition the amount of gas and debris pumped into the atmosphere must have been phenomenal. Somewhat nearer our time, just two million years ago, a mighty eruption at Yellowstone in Wyoming was violent enough to leave behind a gigantic crater (or caldera) up to 80km across, and blast out ash that fell across 16 states. Another huge eruption occurred at Yellowstone around 1.2 million years before the present and yet another just 640,000 years ago.

If this last cataclysm occurred today it would leave the United States and its economy in tatters and the global climate in dire straits. The eruption scoured the surrounding countryside with pyroclastic flows, whose gross volumes were sufficient—if spread across the nation—to cover the entire USA to a depth of 8cm. Ash fell as far afield as what is now El Paso (Texas) and Los Angeles (California), and is even picked up in drill cores from the Caribbean seabed. Although no eruptions have been recorded at Yellowstone for 70,000 years, the hot springs, spectacular geysers, and bubbling mud pools provide testimony that molten rock still resides not far beneath the surface. This is further supported by the numerous earthquakes that regularly shake the region and the periodic swelling and subsiding of the land surface. No one knows when—or even if—Yellowstone will experience another devastating super-eruption. The return periods between the three greatest Yellowstone blasts range from 660,000 to 800,000 years, so we could reasonably expect another sometime soon or have to wait well over a hundred millennia. It is also perfectly possible that Yellowstone may never produce another super-eruption, and that the giant volcanic system will gradually fade away into final extinction.

It would be easy to sit back and say—that's all very well, but such horrific volcanic events occurring on such mammoth scales took place deep within the mists of time. Surely they can't happen today? Thinking along these lines would be a very big mistake. It is true that the likelihood of a super-eruption happening in the lifetime of anyone alive today is small, but at about 714 to 1, not that small. When the next one explodes onto the scene, its impact on modern society could be cataclysmic, both for the host region and the world as a whole. We can get some inkling of what to expect by looking back 74,000 years, to a time when what was perhaps the greatest volcanic explosion of the last couple of million years tore open a hole 100km across at Toba in northern Sumatra (see Figure 13). According to some, the eruption of Toba may have come within a hair's breadth of making the human race extinct. Estimates of the size of the blast vary, but the total amount of debris ejected during the eruption was probably on the order of 3,000 cubic km: sufficient to cover virtually the whole of India with a layer of ash one metre thick.

Any humans living in Sumatra at the time would without question have been obliterated. For the human race as a whole to suffer the threat of extinction, though, the effects of the eruption would have to have been severe across the whole planet, and the balance of evidence suggests that this was the case. Along with the huge quantities of ash, the Toba blast may have poured out enough sulphur gases to create up to 5,000 million tonnes of sulphuric acid aerosols in the stratosphere. This would have been sufficient to cut the amount of sunlight reaching the surface by 90 per cent, leading to global darkness and bitter cold (see Figure 14). Temperatures in tropical regions may have rapidly fallen by up to 15°C, wiping out the sensitive tropical vegetation, while over the planet as a whole the temperature drop is likely to have been around 5°C or 6°C; broadly the equivalent of plunging the planet into full ice age conditions within just a few months and leading to ice and snow covering up to one third of the Earth's surface. Temperature records from Greenland ice cores suggest that the eruption was followed by at least six years of such volcanic winter conditions, which were in turn followed by a thousand-year cold 'snap'. Soon afterwards the planet entered the last Ice Age, and there is some speculation that, in this respect, the cooling effect of the Toba eruption may have been the final straw, tipping an already cooling Earth from an interglacial into a glacial phase from which it only fully emerged around 10,000 years ago.

What then of our unfortunate ancestors: could this period of volcanic darkness and cold really have brought them to their knees? It certainly seems possible. Studies of human DNA contained in the sub-cellular structures known as mitochondria reveal that we are all much too similar—genetically speaking—to have evolved continuously and without impediment for hundreds of thousands of years. The only way to explain this

13. The gigantic eruption of Toba 74,000 years ago excavated a crater 100km long and plunged the world into the depths of volcanic winter

extraordinary similarity is to invoke the occurrence of periodic population 'bottlenecks' during which time the number of human beings was, for one reason or another, slashed and the gene pool dramatically reduced in size. At the end of the bottleneck, all individuals in the rapidly expanding population carry the inherited characteristics of this limited gene pool, eventually across the entire planet. Mike Rampino, a geologist at New York University, and anthropologist Stanley Ambrose of the University of Illinois have proposed that the last human population bottleneck may have been a consequence of the Toba super-eruption. They argue that conditions after the Toba blast would have been comparable to the aftermath of an all-out nuclear war, although without the radiation. As the soot from burning cities and vegetation would result in a 'nuclear winter' following atomic Armageddon, so the billions of tonnes of sulphuric acid in the stratosphere following Toba would mean perpetual darkness and cold for years. Photosynthesis would slow to almost nothing, destroying the food sources of both humans and the animals they fed upon. As the volcanic winter drew on, our ancestors simply starved to death, leaving fewer

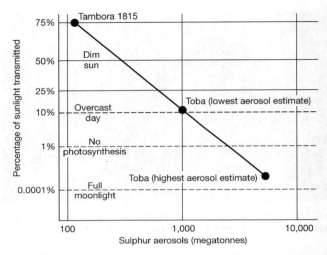

Percentage of sunlight transmitted (y-axis): 75%, 50%, 25%, 10%, 1%, 0.0001%

Tambora 1815

Dim sun

Overcast day

No photosynthesis

Toba (lowest aerosol estimate)

Full moonlight

Toba (highest aerosol estimate)

Sulphur aerosols (megatonnes): 100, 1,000, 10,000

14. Sunlight reduction due to Toba: worst-case estimates suggest that sulphuric acid aerosols from Toba may have cut out so much sunlight that the entire Earth was almost as dark as on a night of full moon

and fewer of their number, perhaps in areas sheltered for geographical or climatological reasons from the worst of the catastrophe. It has been suggested that for 20 millennia or so there may have been only a few thousand individuals on the entire planet. This is just about as close to extinction as a species is likely to get and still bounce back, and—if true—must have placed our ancestors in as vulnerable a position as today's White Rhinos or Giant Pandas.

Could a future super-eruption wipe out the human race? It is highly unlikely that any eruption would be of sufficient size to completely obliterate today's teeming billions, but it is perfectly possible that our global technological society would not survive intact. Before the fall of the Berlin Wall, many national governments were quite prepared to plan for the terrible possibility of all-out nuclear war. With that threat now largely dissipated, however, there has been little enthusiasm for maintaining civil defence plans to address the threat of a global geophysical catastrophe. In the absence of such forward thinking, the impact of a future super-eruption is likely to be severe. With even developed countries such as the United States, the UK, Germany, and Australia having sufficient stores to feed their populations for a month or two at most, how would they cope with perhaps another five or six years without the possibility of replenishment? In the world's poorer countries, where famine and starvation are never far away, the situation would be magnified a thousand times, and death would come swiftly and terribly. From London to Lagos the law of the jungle would quite likely prevail as individuals and families fought for sustenance and survival. When the skies finally cleared and the Sun's initially feeble rays brought the first breath of warmth to the frozen Earth, maybe a quarter of the human population would have died through famine, disease, and civil strife.

It is extremely unlikely, but not impossible, that another super-eruption might strike within the next hundred years. But where? 'Restless' calderas, which are constantly swelling and shaking, are clear candidates, and both Yellowstone and Toba belong in this category. Large volumes of magma still reside beneath these sleeping giants that may well be released in future cataclysms. It was thought likely that the warning signs of these giants' awakening—extensive earthquake swarms and large-scale swelling of the surface—would continue for decades or even centuries before they finally let loose, but new evidence suggests that things might progress far more quickly, so that the period over which warning signs would be apparent could be as little as weeks to months. Notwithstanding recent swarms of earthquake activity at Yellowstone in 2010 and 2013, alongside new data indicating that the magma chamber beneath is twice as big as previously thought, neither this volcano, nor Toba, is displaying such ominous behaviour at the moment that we need to lose sleep over the imminence of a super-eruption. There is some concern,

71

however, over the fact that just a tiny percentage of the 1,500 or so active volcanoes that have erupted in the last 10,000 years are being monitored, so we have no idea what is going on deep beneath most of them. Furthermore, the next super-eruption may blast itself to the surface at a point where no volcano currently exists. Even as I write this, perhaps, some gigantic mass of magma that has been accumulating deep under the remote southern Andes—for example—may be priming itself to tear the crust apart, and our familiar world with it.

The super-eruptions I have talked about so far have all been cataclysmically explosive affairs. There is, however, another much less common species. One that, every few tens of millions of years, erupts even greater volumes of magma, but with relatively little violence. Flood basalt eruptions involve the effusion of prodigious volumes of low-viscosity lava that spread out over enormous areas. Such spectacular outpourings are recognized all over the world, including India, southern Africa, the north-west United States, and north-west Scotland, but the greatest breached the surface nearly 250 million years ago in northern Siberia. Estimates vary, but it looks as if the lavas erupted by this unprecedented event covered more than 25 million square km—an area three times that of the United States.

This, and several comparable outpourings during Earth's long history, have been correlated with mass extinctions. Before the Siberian outburst, for example, the Earth of the Permian period teemed with life. During the succeeding Triassic period, however, when the great flows had cooled and solidified, fully 95 per cent of all species had vanished from the face of the planet. A similar mass extinction 65 million years ago, at the end of the Cretaceous period, has been linked to the enormous Deccan Trap flood basalt eruption in north-western India. As I will address in the next chapter, however, there is incontrovertible evidence that the Earth was struck at this time by a comet or asteroid, and, although the debate continues, the consensus has it that this was the primary cause of the extinction of the dinosaurs and numerous other species at the end of the Cretaceous. Nonetheless, the Deccan lavas may also have had a role to play, pumping out colossal quantities of carbon dioxide that may have led to severe greenhouse warming and the demise of any organisms that were unable to adapt quickly enough. As our polluting society continues to do the same, perhaps we should take this as a salutary warning of what the future might hold for us, our world, and life upon it.

A watery grave

Although by no means the largest volcanic event of the 20th century, the spectacular 1980 eruption of Washington State's Mount St Helens volcano proved to be a watershed, because it drew attention to a style of eruption that had previously attracted little interest from volcanologists. Most eruptions involve the vertical ejection of volcanic debris from a central vent, but the climactic eruption of Mount St Helens was quite different. Lava and debris from the previous eruption—all of 120 years earlier—had blocked the central conduit ensuring that the fresh magma rising into the volcano could not easily escape. Instead it forced its way into the volcano's north flank, causing it to swell like a giant carbuncle. By mid-May 1980, the carbuncle was 2km across and 100m high, and very unstable. Just after 8.30 in the morning on 18 May, a moderate earthquake beneath Mount St Helens caused it to shrug off the bulge, which within seconds broke up and crashed down the flank of the volcano as a gigantic landslide. With this huge weight removed from the underlying magma, the gases contained therein decompressed explosively, blasting northwards with sufficient force to flatten fully grown fir trees up to 20 km away and obliterating, in all, over 600 square km of forest. The landslide material mixed rapidly with river and lake water forming raging mudflows that poured down the river valleys draining the volcano, while pyroclastic flows covered the flanks and ash fell as far afield as Montana 1,000km away.

The Mount St Helens blast killed 57 people and was a disaster for the region, but its scientific importance lies squarely in its elucidation of the mechanism known as volcano lateral collapse. Most of us view volcanoes as static sentinels: bastions of strength and rigidity that are unmoving and unmovable, when in reality they are dynamic structures that are constantly shifting and changing. Far from being strong they are often rotten to the core: little more than unstable piles of ash and lava rubble looking for an excuse to fall apart. The numerous studies that followed the Mount St Helens eruption revealed that collapse of the flanks and the formation of giant landslides is a normal part of the life-cycle of many volcanoes, and probably occurs somewhere on the planet at least half a dozen times a century. Furthermore, they showed that the Mount St Helens landslide was tiny compared to the greatest known volcano collapse, having a volume of less than a cubic km compared with over 1,000 cubic km for the prodigious chunks of rock that have, in prehistoric times, sloughed off the Hawaiian Island volcanoes.

At this stage you might be asking yourself, so what? Surely a hunk of rock—however large—falling off a volcano can't have a global impact—can it? Well it probably can, provided that the collapse occurs into the ocean. In 1792, a relatively small landslide flowed down the side of Japan's Unzen volcano and into the sea. The water displaced formed tsunamis tens of metres high that scoured the surrounding coastline, killing over 14,000 people in the small fishing villages that lined the shore. Just over a century later, in 1888, part of the Ritter Island volcano, off the island of New Britain (Papua New Guinea), fell into the sea, generating tsunamis up to 15m high that crashed into settlements on neighbouring coastlines, taking over 3,000 lives. Clearly, the combination of a volcanic landslide and a large, adjacent, mass of water is a lethal one, but—you are no doubt thinking—how can it affect the vast majority of the Earth's population who live far from an active volcano? The answer lies partly in the size of the largest collapses and partly in the scale of the tsunamis they generate.

Underwater images of the seabed surrounding the Hawaiian Islands show that they are surrounded by huge aprons of debris shed from their volcanoes over tens of millions of years. Within this great jumbled mass of volcanic cast-offs, nearly 70 individual giant landslides have been identified, some with volumes in excess of 1,000 cubic km. The last massive collapse in the Hawaiian Islands occurred around 120,000 years ago from the flanks of the Mauna Loa volcano on the Big Island. Giant tsunamis resulting from the entry of this huge mass of rock into the Pacific Ocean surged 400m up the flanks of the neighbouring Kohala volcano—higher than New York's Empire State Building. Deposits of a similar age, which may be tsunamis-related, have also been recognized 15m above sea level and 7,000km away on the southern coast of New South Wales in Australia. While the nature and provenance of the latter deposits is still debated, the scale of the waves generated appears to be realistic, and computer models developed to simulate the emplacement of giant volcanic landslides into the ocean come up with similar-sized tsunamis.

It seems, then, as if major collapses at ocean island volcanoes are perfectly capable of producing waves that are locally hundreds of metres high and remain tens of metres high even when they hit land half an ocean away. The next collapse in the Hawaiian Islands is likely, therefore, to generate a series of giant tsunamis that will devastate the entire Pacific Rim, including many of the world's great cities in the United States, Canada, Japan, and China. In deep water, tsunamis travel with velocities comparable to a Jumbo Jet, so barely 5–8 hours will elapse before the towering waves crash with the force of countless atomic bombs onto the coastlines of North America and eastern Asia.

Nor is the problem confined solely to the Pacific. Scientific cruises around the Canary Islands, together with detailed geological surveys on land, have revealed a picture very similar to that painted for Hawaii. Huge masses of jumbled rock stretching for hundreds of kilometres across the seabed, and gigantic cliff-bounded collapse scars on land, testify to prodigious prehistoric collapses from the islands of Tenerife and El Hierro. Of more immediate concern, it looks as if a new giant landslide has recently become activated on La Palma, the westernmost of the islands, and is primed and ready to go. During the eruption before last, in 1949, a large chunk of the western flank of the island's steep and rapidly growing volcano—the Cumbre Vieja—dropped up to 4m towards the North Atlantic and then stopped. Recent monitoring by my own team of scientists from UCL, using the Global Positioning System, reveals that this gigantic chunk of volcanic rock—with an estimated volume of maybe 500 cubic km, just about double the size of the UK's Isle of Man—is now detached from the main body of the volcano and moving independently westwards at the rate of a centimetre or so a year. It might seem that movement at such a snail's pace would be nothing to worry about, but if events follow the expected path, this stupendous volume of creeping volcanic rock will ultimately crash *en masse* into the sea.

The problem at the moment is that we don't have a clue when this will happen. It will probably be soon—geologically speaking—but whether it will be next year or in 10,000 years we simply don't know. We strongly suspect, however, that some sort of extra shove will be needed to translate the tiny lateral movements that we observe today into a careering landslide of colossal dimensions, travelling as fast as a Formula 1 racing car. Most probably, then, the west flank of the Cumbre Vieja will complete its journey to the sea during a future eruption, when a combination of the push of fresh magma and the seismic shaking associated with its ascent and eruption will provide the perfect conditions for collapse. Whether during the next eruption, or fifty eruptions down the line, this will happen, launching a tsunami that will bring devastation to the coastlines of the Canary Islands and perhaps much further afield. Steve Ward of the University of California at Santa Cruz and Simon Day of University College London's Aon Benfield UCL Hazard Centre created quite a stir in 2001 when they published a scientific paper that modelled the future collapse of the Cumbre Vieja, predicting that the resulting tsunami would not only be regionally catastrophic, but would be sufficiently powerful to ravage the entire North Atlantic Basin (see Figure 15). Within two minutes of the landslide entering the sea, Ward and Day show that—for a worst-case scenario involving the collapse of 500 cubic km of rock—an initial dome of water an almost unbelievable 900m high will be generated, although its height will rapidly diminish. Over the next 45 minutes a series of gigantic waves up to 100m high will pound the shores of the Canary

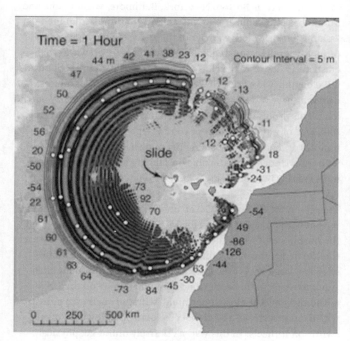

15. A model of the worst-case collapse of the Cumbre Vieja predicts the generation of colossal tsunamis that will batter Africa, Europe, and the Americas. Wave-crest heights are indicated by positive numbers and troughs by negative numbers (all heights in metres). The map shows the position of the waves one hour after collapse

Islands, obliterating the densely inhabited coastal strips, before crashing onto the African mainland. As the waves head further north they will start to break down, but Spain and the UK will still be battered by tsunamis up to 7m high. Meanwhile, to the west of La Palma, a great train of huge waves will streak towards the Americas. Barely six hours after the landslide, waves tens of metres high will inundate the north coast of Brazil, and a few hours later pour across the low-lying islands of the Caribbean and impact all down the east coast of the United States. Focusing effects in bays, estuaries, and harbours could increase wave heights to 50m or more as Boston, New York, Baltimore, Washington, and Miami bear the full brunt of Vulcan and Neptune's combined assault. The destructive power of these skyscraper-high waves cannot be underestimated. Unlike the wind-driven waves that crash every day onto beaches around the world, and which have wavelengths (wave crest to wave crest) of a few tens of metres, tsunamis have wavelengths that are typically hundreds of *kilometres* long. This means that once a tsunami hits the coast as a towering, solid wall of water, it just keeps coming—perhaps for five or ten minutes or more—before taking the same length of time to withdraw. Under such a terrible onslaught all life and all but the most sturdily built structures would be obliterated.

The lessons of the Indian Ocean and Japan tsunamis have taught us that without considerable forward planning it is unlikely that the nine hours it would take for the wave's advance guard to reach the North American coastline would be sufficient to facilitate effective, large-scale evacuation, and the death toll would likely run into many millions. Furthermore, the impact on the US economy would be close to terminal, with the insurance industry wiped out at a stroke and global economic meltdown following swiftly on its heels. In this way, a relatively minor geophysical event at a remote Atlantic volcano would affect everyone on the planet. Like volcanic super-eruptions, these giant tsunamis constitute perfectly normal, albeit infrequent, natural phenomena. At some point in the future one will certainly wreak havoc in the

Atlantic or Pacific Basins, but when? The frequency of collapses on the Hawaiian volcanoes has variously been estimated to be between 25,000 and 100,000 years, but if giant landslides at all volcanic islands are considered, it may be that a major collapse event occurs every ten millennia or so. On a geological timescale this is very frequent indeed and should provide us with serious cause for concern. Even more worryingly, the rate of collapse may not be constant and the contemporary warming may in fact bring forward the timing of the next collapse. My own research team has linked increased incidences of past volcano collapse with periods of changing sea level, while others have suggested that a warmer and wetter climate might result in greater numbers of large volcanic landslides. Given that sea levels are forecast to continue to rise for the foreseeable future, while studies of past climate change show that a warmer planet results in heavier rainfall on many of the world's largest volcanic island chains, perhaps we should all be thinking of moving inland and uphill, or at least of investing in a good-quality wet suit.

The city waiting to die

It is extraordinarily difficult to get across to someone who has never experienced it, the sheer, mind-numbing terror of being caught in a major earthquake. Even in California, where the population is constantly bombarded with information about what to do in the event of a quake, coherent, logical thought ceases when the ground starts to tremble. Following the Loma Prieta earthquake that struck northern California in 1989, a survey by the United States Geological Survey revealed that only 13 per cent of the population of Santa Cruz sought immediate protection, while close to 70 per cent either froze or ran outside. This is a perennial problem with earthquakes; however well informed they are, when the ground starts bucking like a bronco and the furniture starts to hurl itself across the room, people's blind instinct takes over and tells them to 'get the hell out of there'. Unfortunately, this serves only to increase the death toll as terrified home-owners rushing screaming into the street provide easy targets for falling masonry and other debris crashing down from above. What they should be doing is diving beneath the nearest piece of heavy furniture or sheltering beneath the lintel of a convenient doorway.

Earthquakes are immensely destructive, mainly because most cities in regions of high seismic risk are dominated by buildings that are simply not built well enough to withstand the severe ground shaking of a big quake. Modern construction methods in California follow stringent building codes that ensure they can withstand earthquakes that would be devastating elsewhere, and this policy has borne considerable fruit by dramatically limiting death, injury and damage during major quakes in the last quarter of a century. Even so, the Northridge earthquake that struck southern California in 1994 is credited with losses totalling US$35 billion, largely accruing from damage to older structures. Other quake-prone countries also have in place building codes designed to minimize damage due to ground shaking, but often these codes are inadequate or simply not enforced. The terrible legacy of such a lack of commitment by government and local authorities became all too apparent when a magnitude 7.4 quake struck the Izmit region of Turkey in 1999, obliterating 150,000 buildings and taking more than 17,000 lives. Many apartment blocks simply 'pancaked', successive floors collapsing to form a stack of concrete slabs beneath which opportunities for survival were minimal. In January 2001, a severe earthquake shook the Bhuj region of Gujarat state in north-western India, flattening 400,000 homes and killing perhaps 100,000 people. Many of the deaths resulted from the traditional construction methods used in the region, which involved the building of homes with enormously thick walls made of great boulders held together loosely with mud or cement, beneath heavy stone roofs. When the ground started to shake these buildings offered little resistance, collapsing readily to crush those inside. Two years on, a moderate earthquake in southern Iran took 26,000 lives in the city of Bam, as the traditional adobe (mud brick) buildings put up little or no resistance to the ground shaking. And the list goes on, with poor building standards almost entirely responsible for the combined total of more than one-third of a million lives lost in the Kashmir, Pakistan (2005), Sichuan, China (2008), and Haiti (2010) quakes.

During the last millennium, earthquakes were responsible for the deaths of at least eight million people. Terrifying as this sounds, the rapid growth of megacities in regions of high seismic risk is set to ensure that this figure is surpassed, maybe in just the next few centuries, and some seismologists are already warning of the potential, in the near future, for a single large quake to take 3 million lives. If the unfortunate target were Karachi or Mexico City, although the catastrophe would have appalling consequences for the host countries, the global impact would be minimal and would barely impinge upon the lives of most of the world's population. On the other hand, if ground zero were to be the Japanese capital, Tokyo, then the story would be very different. With a population a shade over 37 million, the Tokyo–Yokohama conurbation is the greatest urban concentration on Earth. The city is located in one of the most quake-prone parts of the planet, where the Pacific and Philippine tectonic plates to the east plunge beneath the giant Eurasian plate to the west, and was all but wiped out by a massive earthquake 90 years ago. Things have been ominously quiet since, and it can't be long now before another huge quake devastates one of the world's great industrial powerhouses. When it does, the economic shock waves will hurtle out across the planet. In order to provide an impression of the fate awaiting the Japanese capital, let me take you on a trip back to one of the great disasters of the 20th century: the terrible event the Japanese call the Great Kanto Earthquake.

1 September 1923 dawned like any other day for the inhabitants of Tokyo and Yokohama, but for many it would be their last. The quake struck just before noon, when the cafés and beer halls were packed with hungry and thirsty workers and as families sat down at home to their midday meal. A low, deep rumbling grew rapidly

81

16. Following the Great Kanto Earthquake of 1923, little of Tokyo remained standing after the post-quake fires had raged for two days across the city

to a monstrous roar as a fault below Sagami Bay ripped itself apart and sent shock waves tearing northwards towards the twin cities, crashing first into Yokohama and then—a bare 40 seconds later— into the heart of the capital itself. The quake registered a massive 8.3 on the Richter Scale, and so severe was the ground shaking that it was impossible even to stand. Within seconds, thousands of buildings, many with the traditional heavy tiled roofs, collapsed into heaps of rubble, bringing sudden oblivion to those inside. The great cacophony of grinding rock and falling buildings eventually gave way to the quieter but equally terrifying crackle of flames as fires started by thousands of overturned stoves began to devour the wood out of which many of the buildings were constructed. Whipped up by a brisk wind, a million small fires swiftly merged to form unstoppable walls of flame that marched across the ruins. Shocked men, women, and children cowered before them in open spaces, but to no avail. The firestorms roasted them alive. In one area of waste ground 40,000 were immolated by the conflagration, so packed together that some charcoaled bodies were found still upright. The fires continued to consume what remained of the cities for two days and nights, before finally burning themselves out to reveal a post-apocalyptic scene of utter devastation (see Figure 16). The true total will never be known but up to 200,000 people may have lost their lives in the quake itself and the fires that followed. The cost to the Japanese economy was phenomenal—around US$60 billion at today's prices—and a combination of the quake and the Great Depression six years later led to economic collapse and severe hardship. Some have even suggested that these circumstances, as in the German Weimar Republic, helped stoke the fires of nationalism and the rise of the military, leading to imperialism, conquest, and ultimately war.

Once again, in the second decade of the new millennium, the twin cities of Tokyo and Yokohama together await their fate, only this time it will be far, far worse, both for Japan and the rest of the world. Now the industrial and commercial might of the region constitutes one of the major hubs of the world market, with spokes reaching out to the far corners of the Earth, helping to bind together a global economic machine upon which the wealth of all nations now depends. When Tokyo falls, so will Japan, and the rest could follow—but when? Strains have now been accumulating in the rock beneath and around the capital for nine decades, with no major earthquake to ease the accumulating pressure. Despite being rocked in 1992 by a magnitude 5.7 event located about 30km to the south, which resulted in minor damage, and by other quakes in the region that have caused buildings to sway, the bedrock beneath Tokyo has remained largely seismically silent for the best part of a century. Both the government and the population know, however, that this can't last and money is being poured into constructing earthquake-proof buildings, emergency planning, and training the population; even trying to predict the precise timing of the next 'big one'. So far, however, the accurate prediction of earthquakes has proved to be out of reach, and prospects for a breakthrough in the near future are slim. Furthermore, a substantial proportion of the older building stock remains vulnerable, and an estimated one million wooden buildings continue to provide an excellent potential source of fuel for post-quake fires.

Eighteen years ago and 400km south of Tokyo, 6,000 people died in the Kobe earthquake, which can perhaps be viewed as a mini version of the catastrophe awaiting the capital. At Kobe serious fire damage contributed significantly to the overall destruction and to the huge economic losses of more than US$150 billion. The event also made clear the fact that emergency preparedness and response were far from effective, and certainly well below the rest of the world's expectations, given the general perception of Japanese society as a model of efficiency. For one reason or another, the authorities were simply unable to cope with the chaotic aftermath of the event. Plans were not in place to ensure transport of emergency supplies and equipment to where they were needed, once roads were blocked by debris and railways out of commission, and many of the city's hundreds of thousands of homeless received little or no help for several days after the quake. It is fair to say that some at least of the problems encountered at Kobe reflect the hierarchical structure of Japanese society, which has a tendency to stifle independent decision making and action and hinders rapid response in emergency situations. Without significant changes it is difficult to see how any earthquake emergency plan for the Tokyo region could function effectively within the straitjacket imposed by such a deeply ingrained and restrictive social etiquette.

The geological setting of Tokyo and Yokohama is complex, with three of the Earth's great tectonic plates converging here. The enormous strains associated with the relative movements of these plates are periodically relieved by sudden displacements along local faults, which in turn lead to destructive earthquakes. In fact, there are so many active faults in the vicinity that a number of major earthquake scenarios are envisaged. While a repeat of the

Great Kanto Earthquake may be a century or so away, a so-called chokka-gata quake, occurring directly beneath the capital, is the most feared in the near future. Although smaller than the 1923 event, at around magnitude 7, its proximity would result in massive damage to Tokyo. Worryingly, the capital is under much greater threat today than it was prior to March 2011. This is because the huge magnitude 9.0 Tohuku quake, 370km to the north, not only triggered a catastrophic tsunami but also piled additional strain onto the faults in the Tokyo area, making their rupture more likely in the shorter term, thereby increasing the seismic hazard for the capital by two or three times. So primed are the faults beneath Tokyo that a 2012 study by Japanese scientists suggested that the probability of a magnitude 7+ quake occurring beneath Tokyo could be as high as 70 per cent by 2016, and 98 per cent within the next 30 years. In other words, a shoo-in cert.

The national government still maintains that its scientists will detect in advance the warning signs that the 'big one' is on its way. Such faith in science is both rare and touching, but in this case entirely misplaced. Retrospectively, it has been noted that some earthquakes have been preceded by falls in the water levels in wells and boreholes, and in elevated concentrations of radioactive radon gas issuing from the rock, but this is not always observed. Furthermore, such changes can occur without a following quake, making them notoriously unreliable for prediction purposes. A group of Greek scientists claim that they can detect electrical signals in the crust prior to an earthquake, but there is no convincing evidence that this can be used effectively as a predictive tool. On the other hand, there does appear to be something in the idea that animals, birds, and fish behave strangely before an earthquake, and the Japanese are actually undertaking serious research to find out if catfish—amongst other organisms—can help them forecast the next big one. The problem here is that no one knows how animals can detect a quake before it happens, although it has been speculated that strain in the rocks generates electrical charges in fur and feathers, and perhaps even scales, that trigger small electric shocks, making the animals understandably restless and irritable. But this begs the question, just how *do* you quantify whether a pig, for example, is behaving strangely?

In the absence of an alert from a pre-cognizant catfish, it is likely that the next big quake will strike the Tokyo region with no warning whatsoever. Recently constructed buildings will fare pretty well, but many older properties will crumble. Notwithstanding automatic gas shut-off devices that are fitted to some buildings, exploding fuel tanks, fractured gas mains, and oil and chemical spills will ensure no shortage of fires to feed on a million wooden buildings. As in 1923, huge conflagrations are expected to cause at least as much destruction as the quake itself, and to inflate substantially the likely death toll—which is expected to be upwards of 20,000 for a magnitude 7.3 quake under Tokyo Bay. While it is difficult to estimate in advance the economic losses resulting from the next Tokyo quake, a magnitude 7 event is forecast to result in losses of at least US$1 trillion. Even worse, a modelling company that services the insurance industry has come up with the extraordinary figure of up to US$4.3 trillion for a repeat of the 1923 Great Kanto Earthquake. This is more than 17 times the cost of the 2011 Tohuku quake and tsunami, to date the world's most expensive natural catastrophe.

The impact on the Japanese economy of the next 'big one' is widely expected to be shattering. Japan is enormously centralized, and the Tokyo region hosts not only the national government but also the stock market and 70 per cent of the headquarters of the country's largest corporations. Japan has the third largest economy on the planet, accounting for around one-third of Asia's GDP and 8 per cent of the world's GDP. Despite ever-fluctuating economic circumstances, it remains pretty certain that the country will still be an economic powerhouse when the next big quake strikes. In order to rebuild and regenerate it is highly likely that the Japanese will have to disinvest from abroad on a massive scale, dumping government bonds in Europe and the States, selling foreign assets, and shutting down overseas factories. It is well within the realms of possibility that as country after country finds itself fighting to cope with the swift unravelling of the global economy, a crash deeper than anything since 1929 or 2008 would soon set in. No one knows how long a post-Tokyo quake depression would last, nor just how bad it would be. Equally importantly, we don't know long we have to wait until such a speculative scenario is played out for real. Perhaps only a few years, perhaps another hundred years or more, but it would be no real surprise if this great city was brought once again to its knees before the next century dawns.

Despite occasionally being depicted in the media as 'Disasterman', I would hate you to regard me purely as a harbinger of doom, and close the book at this point with a feeling of hopelessness about the future. Yes, the Earth is geologically very dangerous, and the more geologists study our planet the more potentially serious the tectonic threat to the survival of our civilization appears to be. On the other hand, we are learning all the time, collecting data that can be utilized to counter or at least mitigate the impact of the next super-eruption or mega-tsunami. Eventually, it might well be possible to predict earthquakes with some accuracy and precision and certainly, within a century, it would be nice to think that nowhere on the planet could a volcanic island become unstable or a huge new batch of magma swell the surface without our satellites spotting them well in advance of catastrophe. On an almost daily basis Earth scientists are tackling some of the greatest threats to our society and incrementally they are getting to grips with them. At the very least, the next time our planet shudders on a grand scale we will be far better prepared than our distant ancestors, who faced the might of the Toba super-eruption with incomprehension and sheer terror.

Chapter 5
The threat from space: asteroid and comet impacts

The astronomical event of the century

In 1993 a discovery by Carolyn Shoemaker, wife of the late and greatly lamented planetary scientist Eugene Shoemaker, and colleague David Levy, was to change for ever our perception of the Earth as a safe and cosy haven insulated from the whizzes and bangs of a violent and capricious universe. The Shoemaker team had spotted 21 huge chunks of rock that had once been part of a comet torn apart by the enormous gravitational field of the planet Jupiter: a giant sphere mainly made up of hydrogen and helium gas that is large enough to contain over 1,300 Earths. Instead of orbiting the Sun, like most comets, however, this one had been captured by Jupiter's gravity and the rocky fragments now orbited the King of Planets himself. As Jupiter already had a large retinue of 67 moons, the addition of a few more would have been mildly interesting, if not surprising. What was extraordinary, however, was that these new 'moons' were to prove short-lived. The following year they would end their lives by crashing into the surface of Jupiter, providing scientists on Earth with a grandstand view of just what happens when a planet is struck by large hunks of space debris (see Figure 17).

On 16 July 1994—appropriately the 25th anniversary of the launch of Apollo 11, the first manned lunar landing mission—the first

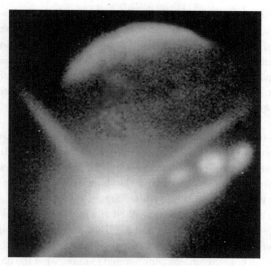

17. The dramatic impact of one of the fragments of Comet Shoemaker–Levy on Jupiter in 1994

fragment of Comet Shoemaker–Levy struck the planet, sending up a gigantic plume of gas and debris and blasting outwards a rapidly expanding shock wave. As fragment after fragment hammered into the planet, spectacular images were gathered by the Hubble Space Telescope in Earth orbit and by the unmanned Galileo probe on its way to Jupiter. Two days after the initial impact, a chunk of rock 4km across and rather unromantically named fragment G smashed into the planet with the force of 100 million million tonnes of TNT—roughly the equivalent of eight *billion* Hiroshima-sized atomic bombs. The flash generated by the collision was so brilliant that many infra-red telescopes trained on the event were temporarily blinded. The glare soon faded, however, to reveal an enormous dark impact scar wider than the Earth. Inevitably, everyone who saw this awesome image had the same thought. What would have happened if fragment G had struck the Earth instead of Jupiter? Almost overnight our planet

89

seemed a much more vulnerable place and the hold of our race upon it that much more tenuous. Suddenly both scientists and the public, and even politicians, began to take the threat from space seriously. Two Hollywood blockbuster films fed growing interest in impact events by showing—with various degrees of scientific rigour—what we might all be in for if a comet or asteroid headed our way.

In 1996, just two years after the Jupiter impacts, an international body known as the Spaceguard Foundation was formed, with the dedicated aims of promoting the search for potentially dangerous asteroids and comets and raising the general level of awareness of the impact threat. In the United States, NASA and the Department of Defense began, albeit at a low level, to fund Spaceguard-related projects, and the UK government established a task force to examine the risk of asteroids and comets hitting the Earth. All of a sudden everyone wanted to know what the chances were of the Earth being struck at some point in the future and what effect such a collision would have on our planet and our race. The answer to the first question is easy: the chances are 100 per cent. The Earth has been bombarded by space debris throughout its long history, and although such collisions are now far less common than they were billions of years ago, our planet will be struck again. The vital question is—when? And as regards how bad this will be for the human race: it depends largely upon how big a chunk of rock hits us.

The cosmic sandstorm

To get a better idea of how frequently the Earth is likely to be hit, we need to find out how many rocks are hurtling around our solar system and, in particular, how many of these come close enough to the Earth to start us worrying. Although a vast amount of debris was swept up by the embryonic planets during the early history of the solar system, countless dregs remain, ranging upwards in size from tiny specks a few millimetres across to gigantic rocks, such as the minor planet Ceres, more than 1,000km in diameter. Like someone battling through a desert sandstorm, the Earth is constantly bombarded in the course of its journey through space. Fortunately for us, most of the billions of colliding fragments are tiny and flash into oblivion as soon as they come into contact with our planet's atmospheric shield. Every now and again, however, the Earth collides with something larger.

A fragment of debris the size of a pea burns up in the Earth's atmosphere every five minutes, while a soccer-ball-sized lump will light up the sky with its death throes around once a month. Larger objects may run the gauntlet of the atmosphere and reach the surface, but this is rare and only happens a few times a year. Perhaps every few centuries, the Earth collides with a rock in the 40–50m size range—an object large enough to obliterate a city if it scores a direct hit. The last well-documented impact of this size occurred as recently as 1908—of which more later.

While the entire solar system teems with debris, from a hazard point of view we are only really interested in those fragments that threaten to end their existence through collision with our world. The majority of these Earth-threatening objects are rocky asteroids that have orbits around the Sun that approach or intersect the Earth's. The true numbers of these Near Earth Asteroids (NEAs) are impossible to determine, but current estimates are pretty frightening (see Figure 18). In all, up to 20 million pieces of rock over 10m wide may be hurtling across or coming close to our planet's path during its journey around the Sun. Up to 100,000 of these are thought to be over 100m in diameter—big enough to obliterate London or New York given a direct hit—and maybe 20,000 are half a kilometre across, sufficient to wipe out a small country if they strike land, or generate devastating tsunamis if they impact in the ocean. Fewer in number, but enormously more destructive if they hit, are those asteroids 1km or more in diameter, which have the potential to obliterate a country the size of England and—if 2km or more across—wreak havoc across the

91

globe. Although barely equivalent in diameter to twenty soccer pitches laid end to end, such is the tremendous level of kinetic energy (energy of motion) involved in the collision that a 2km object striking land would leave a crater 40km or so across and loft sufficient pulverized debris into the atmosphere to block out the Sun's rays and plunge the Earth into a freezing *cosmic winter* for years.

A range of estimates have been published for the number of NEAs in the 1km and above size range, with the most recent suggesting there are around one thousand. As of September 2013, 859 of these had been identified, perhaps 85 per cent of the total, and their orbits projected forward in time to see if they pose a threat to the Earth in the medium term. The search continues to find them all—a task that will take at least a couple more decades. Once this has been accomplished, and assuming that one does not have our name on it, we can sleep a little safer in our beds. The problem does not, unfortunately, end there. We still have the comets to worry about.

Comets are enormous masses of rock and ice that can be up to 100km or more across. In contrast to the near-circular orbits of the asteroids, most comets follow strongly elliptical paths that carry them from the frigid outposts of the outer solar system, or beyond, in close to the Sun and then out again. In the depths of space, comets are enigmatic objects and not easy to spot. As they enter the inner solar system, however, they undergo a remarkable transformation as sunlight starts to evaporate gas and dust particles from the central nucleus to form a spectacular 'tail' that can stretch across space for 100 million km or more. The stunning 'apparition' of a comet's tail was long regarded as a portent of doom and disaster, and in a way this is not too far from the mark. Comets have typical velocities of 60–70km a second: a hundred times faster than the Concorde supersonic airliner, and around three times that of the typical NEA. As a result, a collision between a comet and the Earth would be hugely more energetic and therefore tremendously more destructive.

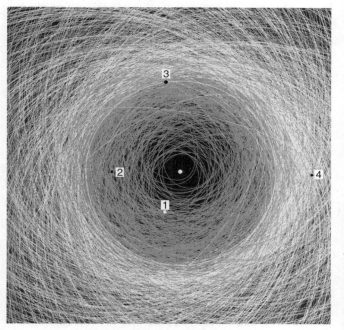

18. Orbits of known Earth-crossing asteroids provide a clear picture of just how crowded the space around our planet really is. The orbits of 1. Mercury 2. Venus 3. Earth 4. Mars are also shown

Another problem with comets is that, unlike their asteroid cousins, their orbital parameters are often poorly known and therefore difficult to project into the future to see if they pose any threat. Halley's Comet, undoubtedly the most famous of all, follows an orbit around the Sun that takes only 76 years to complete. Consequently, it has been observed dozens of times over thousands of years and its orbit is well enough known to make it possible to calculate its path far into the future. This shows that, at least until AD 3000, Halley's Comet will not even come close to threatening the Earth. Other comets, however, follow parabolic orbits that take them on immeasurably long journeys far beyond the limits of the solar system. Some of these may have been observed once or twice by our distant ancestors, but others may be making their first ever visits to the inner solar system. In these circumstances, there has been no opportunity to predict their orbits on the basis of earlier visitations, and our first view of one of these objects heading our way may provide us with just six months respite before an unavoidable and calamitous collision. Furthermore, because such comets have been confined to deep space, they are huge—perhaps 100km or more across. This is because they have not suffered attrition from the solar wind—the hurricane of solar particles that evaporates parts of a comet to produce the characteristic tail, as it forges its way through the inner solar system.

When worlds collide

It has been something of a struggle over the last few centuries for exponents of the theory to convince both other scientists and the public that the millions of craters that pockmark the face of the Moon are the result not of volcanic explosions but of collisions with objects from space. As long ago as the early 19th century, the German natural philosopher Baron Franz von Paula Gruithuisen's declaration that the lunar craters were a consequence of 'a cosmic bombardment in past ages' was treated with contempt by 'serious' scientists (no doubt his further claims to have uncovered evidence for the existence of humans and animals on the surface of the Moon had a little to do with this). At the other end of the 19th century, the US geologist Grove Karl Gilbert tried to simulate in the laboratory the formation of the lunar craters by firing objects into powder or mud. Gilbert was perplexed, however, by the observation that only objects fired vertically produced circular craters like those that cover the lunar surface. In light of this, W. M. Smart proclaimed, in 1927, that the craters of the Moon could not be caused by impacts because 'there is no a priori reason

94

why meteors should always fall vertically'. It was only after observing the effects of the billions of tonnes of bombs dropped in the Second World War that it began to dawn on geologists that given a violent enough explosion, a circular crater would always be formed, whatever the angle of the trajectory. Remarkably, it took another quarter of a century for the impact origin of lunar craters to gain widespread acceptance, but getting any new paradigm accepted in science is a battle, and geology is no exception. Just as the proponents of the revolutionary theory of plate tectonics had initially to fight hard against reactionary forces, so those scientists who claimed that the Earth, as well as the Moon, had also taken a battering found the going difficult.

As long ago as 1905, Benjamin Tilghman proposed that Arizona's famous Barringer Crater (also now known as Meteor Crater) was the result of 'the impact of a meteor of enormous and hitherto unprecedented size'. This suggestion failed to convince, however, because a quarter of a century of excavation by Tilghman and his engineer colleague D. M. Barringer failed to find the impactor itself. We now know that this had been essentially vaporized by the enormous heat generated by the collision, but at the time the absence of a 'smoking gun' simply lent credence to those who suggested an alternative mechanism of formation.

Until well after the end of the Second World War, many Earth scientists suffered an extraordinary failure of the imagination, accepting an impact origin for the lunar craters but grabbing at any straw in order not to support an impact origin for crater structures on the surface of our own world. Given that, due to the Earth's much greater size and stronger gravity field, it must have been struck perhaps 30 times more frequently than its nearest neighbour, this denial is even more extraordinary. Perhaps not entirely surprising, however, when we consider that the enormously dynamic nature of our planet is far from suited to the preservation of impact craters, particularly those of any great age. Because of plate tectonics, and most especially the process of subduction, through which the basaltic oceanic plates are continuously being consumed within the Earth's hot interior, some two-thirds of the Earth's surface is only a few hundred million years old. Bearing in mind that the most intense phase of bombardment occurred during the first few billion years of our planet's history, evidence for this can now only be found in the ancient hearts of the granite continents that are immune to the subduction process. Because they have succumbed to erosion and weathering, perhaps for aeons, these craters are notoriously difficult to spot. Also, the oldest rocks, which are likely to support the most craters, are in remote areas such as Siberia, northern Canada, and Australia, and some craters are so big that their true form can only be seen from space. Today, satellites have helped in the identification of the 300 or so confirmed or probable impact craters scattered all over the world, and the idea that the Earth is susceptible to bombardment from space is now as accepted as plate tectonics.

Controversy has certainly not gone away, however, and argument continues amongst the scientific community, particularly about the frequency and regularity of impacts and—probably of most interest to the layman—about the effects of the next large impact on our civilization. The question of frequency is far from straightforward and serious disagreement exists between schools of thought that, on the one hand, support a constant flux of impactors and, on the other, advocate so-called impact clustering. Notwithstanding the very heavy bombardment of the Earth's early history, followers of impact 'uniformitarianism' support a strike rate that is uniform and unchanging. This is at variance with rival groups of scientists who are promoting an alternative theory of coherent catastrophism, within which the Earth, for one reason or another, periodically comes under attack from increased numbers of asteroids or comets.

If we are realistically to assess the threat of future impacts to our civilization, then clearly it is vital that we resolve as soon as we can the question of whether the number of collisions is likely to continue at its current rate or whether we have a nasty shock in store somewhere down the line. If the former proves to be correct, we can expect business as usual, meaning a collision with a 50m potential city-destroyer every few centuries or so, a half-km small-country obliterator every several tens of millennia, and a 1km global consequence impact event every 600,000 years. Fortunately for us, extinction level events (ELEs), such as that caused by the 10km monster that ended the reign of the dinosaurs 65 million years ago, appear to happen every 50 to 100 million years, so the chances of one striking the Earth soon are tiny. Nonetheless, based upon the above strike rates, and the efficiency with which impact events can erase millions or even billions of lives at a stroke, your chance (and mine) of being killed during an asteroid or comet strike is 350 times greater than winning the UK lottery. Maybe this scares the wits out of you, but the true situation may actually be worse. If the coherent catastrophists are correct then there are certain times when our planet, and perhaps even our entire solar system, travels through a region of space containing substantially more debris than normal, resulting in a significant increase in impact events at all scales.

A number of theories lay the blame for this periodic increase in Earth-threatening space debris on the episodic disruption of the so-called Oort Cloud, a great spherical conglomeration of billions of comets that envelops the entire solar system far beyond the orbit of Pluto. Typically, comets in the cloud travel along such huge orbits, which take some a quarter of the way to the nearest star, that they rarely visit the inner solar system, and then only in ones and twos. However, if some external influence were to interfere with the cloud, so the thinking goes, hundreds or thousands could have their orbits changed, encouraging them to plunge Sunwards, greatly raising the threat of collision with the planets—including our own. A number of suggestions have been put forward for how the Oort Cloud might be periodically disrupted, including due to the passage through the cloud of the mythical and much sought after 'Planet X', which some scientists think may be orbiting far beyond frozen Pluto and the even more remote planetoid, Sedna, or to a dark and distant stellar companion of our own Sun.

An alternative and intriguing theory, known as the Shiva Hypothesis after the Hindu god of destruction and renewal, has been vigorously promoted by Mike Rampino of New York University and his colleagues, who believe that the great extinctions recognized in the Earth's geological record are the result of major impact events that happen pretty regularly every 26–30 million years. Rampino and his colleagues link this to the orbit of our solar system about the centre of our Milky Way galaxy, an orbit that moves up and down in a wave-like motion. Every 30 million years or so, this undulating path takes the Sun and its offspring through the horizontal plane of our disc-like galaxy, when the gravitational pull from the huge mass of stars at the galaxy's core provides an extra tug. This, say the Rampino school, is sufficient to disturb the orbits of the Oort Cloud comets to an extent sufficient to send an influx of new comets into the heart of the solar system, dramatically raising the frequency of large impacts on the Earth. It is just a few million years now since our system last plunged through the galactic plane, so could a phalanx of comets be heading for us at this very moment? By the time we find out it might very well be too late to do anything.

The Shiva Hypothesis calls for a periodicity operating on truly geological timescales, and for this reason is rarely addressed in discussions of the immediate threat from impact events. Much more relevant to considerations of our own safety and survival, and that of our immediate descendants, is a proposal by UK astronomers Victor Clube and Bill Napier that the Earth is struck by clusters of objects every few *thousand* years, and that our planet took a serious pounding as recently as the Bronze Age, just 4,000 years ago. To find out what might cause such a worryingly recent bombardment we need to return to the Oort Cloud.

Leaving aside disturbance of the cloud due to the passage of the solar system around the galaxy, normality sees a new comet from the cloud every now and again falling in towards the inner solar system, maybe as frequently as every 20,000 years. The newcomer is rapidly 'captured' and torn apart by the strong gravitational fields of either the Sun or Jupiter, forming a ring of debris spread out along its orbit, but concentrated particularly around the position of the original comet itself. A large comet, broken up in this way, can 'seed' the inner solar system with perhaps a million kilometre-sized lumps of rock, dramatically increasing the number of Earth-threatening objects, and significantly raising the chances of our planet being hit. Clube, Napier, and others of this particular coherent catastrophist school propose that the last giant comet from the Oort Cloud entered our solar system towards the end of the last Ice Age, a mere 10,000 years or so ago, breaking up to form a mass of debris known as the Taurid Complex. Every December the Earth passes through part of this debris stream, resulting in the sometimes spectacular light show put on by the Taurid meteor shower, as small rocky fragments and gravel-sized stones burn up in the upper atmosphere. These innocuous bits and pieces only represent the tail end of the Taurid Complex, however, the heart of which contains a 5km wide Earth-crossing comet known as Encke and at least 40 accompanying asteroids any one of which would create global havoc if it struck our planet.

The distribution of debris along the Taurid Complex orbit about the Sun is rather like that of runners in a 10,000m race; while the majority are clustered together in a pack, the rest are dotted here and there around the track. Most years, according to the coherent catastrophists, the Earth's orbit crosses that of the Taurid Complex at a point where there is little debris, resulting in a pre-Christmas spectacle and little else. Every 2,500–3,000 years or so, the Earth passes through the equivalent of the runners' pack and finds itself on the receiving end of a volley of rocky chunks perhaps up to 200–300m across. It has been suggested that just such a bombardment around 4,000 years ago led to the fall of many early civilizations during the third millennium BC. Contemporary accounts of chaos and disaster have been interpreted by some in terms of a succession of impacts, too small to have global consequences, but quite sufficient to cause mayhem in the ancient worlds, through generating destructive atmospheric shock waves, earthquakes, tsunamis, and wildfires. Many urban centres in Europe, Africa, and Asia appear to have collapsed almost simultaneously around 2350 BC, and records abound of flood, fire, quake, and general upheaval. These sometimes fanciful accounts are, of course, open to alternative interpretation, and hard evidence for bombardment from space around this time remains elusive. Having said this, seven impact craters in Australia, Estonia, and Argentina have been allocated ages of 4,000–5,000 years and the search goes on for others. Even more difficult to defend are propositions by some that the collapse of the Roman Empire and the onset of the Dark Ages may somehow have been triggered by increased numbers of impacts when the Earth last passed through the dense part of the Taurid Complex between AD 400 and 600. Hard evidence for these is weak and periods of deteriorated climate attributed to impacts around this time can equally well be explained by large volcanic explosions.

How would you like to die?

If supporters of the Taurid Complex model are to be believed, and I should say now that their views remain very much in the minority amongst advocates of the impact threat, then we may have only another thousand years or so before a series of blinding flashes and crashing sonic booms heralds the arrival of the next batch of fragmented comet. Alternatively, we could face oblivion tomorrow or have to wait 100,000 or more years before a city is obliterated or a thousand millennia before the world is plunged into cosmic winter beneath a cloud of pulverized rock. But whenever the skies next fall, how will it affect us? This will depend upon three things: (i) the size of the object; (ii) how quickly it is travelling; and (iii) whether it hits the land or the ocean. All else being equal, the larger the impactor the more devastating and widespread will be its effects. To reiterate, a body in the 50–100m size range carries enough destructive power to wipe out a major city or a small European country or US state. The level and extent of associated devastation will increase progressively with larger impactors until the critical 2km size is reached. In addition to causing appalling destruction on a regional or sub-continental scale, the arrival of an object of this size will affect the entire planet through engendering a period of dramatic cooling and reduced plant growth. For impactors larger than 2 km the effects on the planet's ecosystems become ever more severe until mass extinctions wipe out a significant percentage of all species. The 10km object that struck the Earth off the Mexican coast at the end of the Cretaceous period, 65 million years ago, not only finished off the dinosaurs but also two-thirds of all species living at the time. Even more disturbingly, there is evidence of a major impact event at the end of the Permian period some 250 million years ago that left fewer than 10 per cent of species alive. In all, at least seven out of 25 major extinctions in the geological record have been linked with evidence for large impacts, although as I mentioned in the previous chapter there is a school of thought that plays down the environmental effects of impact events and prefers to implicate huge outpourings of basalt lava in the great extinctions of the past. In truth, both may have roles to play in the great extinctions that are periodically visited upon terrestrial life.

The destructive potential of a chunk of rock hurtling into the Earth is directly related to the kinetic energy it carries (see Figure 19), and this reflects not only the size of the object but also the velocity of the collision. Because they travel substantially faster, therefore, impacts by so-called long-period comets, whose orbits carry them far out into interstellar space, cause more destruction than either NEAs or local comets that follow orbits confined to the heart of the solar system. Both the nature and scale of devastation also depend upon whether the impactor hits the land or the sea. Two-thirds of our planet's surface is covered by water, so statistically this is where the majority of asteroids and comets strike. In such cases, the amount of pulverized rock hurled into the atmosphere might be reduced, compared to a land collision. This small benefit is likely to be at least partly countered, however, by the formation of giant tsunamis capable of wreaking havoc across an entire ocean basin. Furthermore, the gigantic quantities of water and salt injected into the atmosphere may severely affect the climate and even temporarily wipe out our protective ozone shield. Most of the evidence for the environmental effects of impacts comes from studies of just two events, one small and the other enormous.

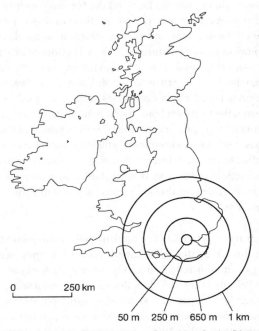

0   250 km

50 m   250 m   650 m   1 km

19. Predicted zones of total destruction for variously sized impacts centred on London

At the low end of the scale, a small asteroid, estimated at around 50m across, penetrated the Earth's atmosphere in 1908 and exploded just 5–10km above the surface of Siberia in a region known as Tunguska. This massive blast, which expended the energy equivalent of somewhere between 800 and 3,200 Hiroshima atomic bombs, flattened more than 2,000 square km of full-grown forest and was heard over an area four times the size of the UK. The blast registered on seismographs thousands of kilometres distant and the atmospheric shock wave was picked up by barographs time and again as it travelled three times around the planet before dissipating. The gas and dust generated by the explosion led to exceptionally bright night skies over Europe, sufficient, according to one contemporary report, to allow cricket to be played in London after midnight. Because of its inaccessibility, the first Russian expedition did not reach Tunguska until a quarter of a century later, when Leonid Kulik and his team were perplexed by the absence of the huge crater they were expecting. Instead they found a circular patch of badly charred and flattened trees 60km across, formed by the airburst as the rock disintegrated explosively due to the huge stresses caused by entry into the atmosphere. As the region was sparsely inhabited, casualties due to the impact were small, with perhaps a few killed and up to 20 injured, although reports are understandably sketchy. Four hours later, however, and the Earth would have rotated sufficiently to bring the great city of St Petersburg into the asteroid's sights and the result would have been catastrophic.

The Tunguska event pales into insignificance when compared to what happened off the coast of Mexico's Yucatan Peninsula 65 million years earlier. Here a 10km asteroid or comet—its exact nature is uncertain—crashed into the sea and changed our world forever. Within microseconds, an unimaginable explosion released as much energy as billions of Hiroshima bombs detonated simultaneously, creating a titanic fireball hotter than the Sun that vaporized the ocean and excavated a crater 180km across in the crust beneath. Shock waves blasted upwards, tearing the atmosphere apart and expelling over a hundred trillion tonnes of molten rock into space, later to fall across the globe. Almost immediately an area bigger than Europe would have been flattened and scoured of virtually all life, while massive earthquakes rocked the planet. The atmosphere would have howled and screamed as 'hypercanes' five times more powerful than the strongest hurricane ripped the landscape apart, joining forces with huge tsunamis to batter coastlines many thousands of kilometres distant.

Even worse was to follow. As the rock blasted into space began to rain down across the entire planet, so the heat generated by its re-entry into the atmosphere irradiated the surface, roasting animals alive as effectively as an oven grill, and starting great conflagrations that laid waste the world's forests and grasslands and turned fully a quarter of all living material to ashes. Even once the atmosphere and oceans had settled down, the crust had stopped shuddering, and the bombardment of debris from space had ceased, more was to come. In the following weeks, smoke and dust in the atmosphere blotted out the Sun and brought temperatures plunging by as much as 15°C. In the growing gloom and bitter cold the surviving plant life wilted and died while those herbivorous dinosaurs that remained slowly starved. Life in the oceans fared little better as poisons from the global wildfires and acid rain from the huge quantities of sulphur injected into the atmosphere from rocks at the site of the impact poured into the oceans, wiping out three-quarters of all marine life. After years of freezing conditions the gloom following the so-called Chicxulub impact would eventually have lifted, only to reveal a terrible Sun blazing through the tatters of an ozone layer torn apart by the chemical action of nitrous oxides concocted in the impact fireball: an ultraviolet spring, hard on the heels of the cosmic winter, that fried many of the remaining species struggling precariously to hang on to life. So enormously was the natural balance of the planet upset that according to some it might have taken hundreds

104

of thousands of years for the post-Chicxulub Earth to return to what passes for normal. When it did the age of the great reptiles was finally over, leaving the field to the primitive mammals—our distant ancestors—and opening up an evolutionary trail that culminated in the rise and rise of the human race. But could we go the same way? To assess the chances, let me look a little more closely at the destructive power of an impact event.

At Tunguska, destruction of the forests resulted partly from the great heat generated by the explosion, but mainly from the blast wave that literally pushed the trees over and flattened them against the ground. The strength of this blast wave depends upon what is called the peak overpressure: that is, the difference between ambient pressure and the pressure of the blast wave. In order to cause severe destruction this needs to exceed 4 pounds per square inch, an overpressure that results from wind speeds that are over twice the force of those found in a typical hurricane. Even though tiny compared with, say, the land area of London, the enormous overpressures generated by a 50m object exploding low overhead would cause damage comparable with the detonation of a very large nuclear device, obliterating almost everything within the UK capital's orbital motorway. Increase the size of the impactor and things get very much worse. An asteroid just 250m across would be sufficiently massive to reach the surface without breaking up, blasting a crater 5km across and devastating an area of around 10,000 square km—an area about the size of the English county of Kent. Raise the size of the asteroid again, to 650m, and the area of devastation increases to 100,000 square km—an area about the size of the US state of South Carolina.

Terrible as this all sounds, however, it would still be insufficient to affect the entire planet. In order to do this, an impactor has to be at least 1.5km across, if it is one of the speedier comets, or 2km in diameter if it is one of the slower asteroids. A collision with one of these objects would generate a blast equivalent to 100,000 *million* tonnes of TNT, which would obliterate an area 500km across—say

105

the size of England—and immediately kill perhaps tens of millions of people, depending upon the location of the impact.

The real problems for the rest of the world would start soon after the impact, as dust in the atmosphere began to darken the skies and reduce the level of sunlight reaching the Earth's surface. By comparison with the huge Chicxulub impact it is certain that this would result in a dramatic lowering of global temperatures but there is no consensus on just how bad this would be. The chances are, however, that an impact of this size would result in appalling weather conditions and crop failures at least as severe as those of the 'year without a summer', which followed the 1815 eruption of Indonesia's Tambora volcano. As mentioned in Chapter 4, with even developed countries holding sufficient food to feed their populations for only a month or so, large-scale crop failures across the planet would undoubtedly have serious implications. Rationing, at the very least, is likely to be the result, with a worst-case scenario seeing widespread disruption of the social and economic fabric of developed nations. In the developing world, where subsistence farming remains very much the norm, widespread failure of the harvests could be expected to translate rapidly into famine on a biblical scale. Some researchers forecast that as many as a quarter of the world's population could succumb to a deteriorating climate following an impact of a 2km object. Anything bigger and photosynthesis stops completely. Once this happens the issue is not how many people will die but whether the human race will survive. One estimate proposes that the impact of an object just 4km across will inject sufficient quantities of dust and debris into the atmosphere to reduce light levels below those required for photosynthesis.

Because we are still not completely certain how many threatening objects there are out there, nor whether they come in bursts, it is almost impossible to say when the Earth will be struck by an asteroid or comet that will bring to an end the world as we know it. Should the Shiva Hypothesis prove to be true, the next swarm of Oort Cloud comets could even now be speeding towards the inner solar system. Failing this, we may have only another thousand years to wait until the return of the dense part of the Taurid Complex and another asteroidal assault. Even if it turns out that there is no coherence in the timing of impact events, there is statistically no reason why we cannot be hit next year by an undiscovered NEA or by a long-period comet that has never before visited the inner solar system. Small impactors on the Tunguska scale pose less of a threat because their destructive footprints are tiny compared to the surface area of the Earth. It would be very bad luck if one of these struck an urban area, and most will fall in the sea. Although this might seem a good thing, a larger object striking the ocean would be very bad news indeed. A 500m rock landing in the Pacific Basin, for example, would generate gigantic tsunamis that would cause massive damage to every coastal city in the hemisphere within 20 hours or so. The chances of this happening are actually quite high—about 1 per cent in the next 100 years—and the death toll could be in the tens of millions if not higher.

Fair enough, you might say, the threat is clearly out there, but is there anything on the immediate horizon? Actually, there is. A dozen or so asteroids, mostly quite small, could feasibly collide with the Earth before 2100. Realistically, however, this is not very likely as the probabilities involved are not much greater than 1 in 10,000—although bear in mind that these are not bad odds. If this was the probability of winning the national lottery then my local agent would be getting considerably more of my business.

In February 2013, we were provided with a spectacular, if relatively benign, reminder of the threat from beyond the atmosphere, when a 10,000 tonne asteroid about twice the size of a London bus hurtled over the Siberian city of Chelyabinsk, before exploding 25km above a sparsely inhabited area with the force of 30 Hiroshima bombs (see Figure 20). The astonishing footage of the fireball and the damage to buildings resulting from the blast, along with the fact that

20. A spectacular fireball and vapour trail mark the passage of a small asteroid above the Siberian city of Chelyabinsk in February 2013

more than a thousand people suffered minor injuries, served to refocus attention on the impact threat. Media concern that the Earth might be the target of some imminent bombardment from deep space was heightened by the close approach of another asteroid, named 2012DA14, which—just 16 hours later—shot past our planet at a distance of 28,000km—closer than many communications satellites. The coincidence of the two events proved, in the end, to be just that, a coincidence, the two chunks of rock following very different orbits.

The two encounters highlight the fact that there are still tens of thousands of objects out there that come close enough to our world to, potentially at least, end their lives here. Looking ahead, however, although some will come close, none of those we know about seem to be zeroing in on the Earth. Best known of the coming threats is the 325m Near Earth Asteroid, MN4, discovered late in 2004 and renamed 99942Apophis, the Greek name for the Egyptian God Apep—the destroyer. At one point, the probability of Apophis striking the Earth on 13 April 2029 was thought to be

108

as high as 1 in 37. Now, to everyone's relief, those odds look to have been wildly pessimistic, and while—like 2012DA14—the rock will barrel past our planet on a track that takes it within the orbits of communications satellites, the probability that it will hit is infinitesimally small. Looking further ahead to 2048, asteroid 2007VK184 is also attracting attention. This 130m rock currently has an impact probability of around 1 in 2,000: a value that ensures the object scores a one on the Torino Impact Hazard Scale, devised to measure the threat from space debris—the only potential impactor to do so at present. There is little concern that 2007VK184 will actually strike the Earth, however, as the Torino Scale wording corresponding to a score of one suggests: 'current calculations show the chance of collision is extremely unlikely with no cause for public attention or concern.' It looks, then, as if we can all sleep soundly—at least for now. Let's hope that many years elapse before we encounter an object that registers a 10 on the Torino Scale, defined as resulting in 'a certain collision with global consequences'. Given sufficient warning, in such circumstances, we might be able to nudge an asteroid out of the Earth's way. Due to its size, high velocity, and sudden appearance, however, there would be little we could do to save ourselves from a new comet with the Earth in its sights.

Epilogue

Now it's been ten thousand years
Man has cried a billion tears
For what he never knew
Now man's reign is through

'In the year 2525': Zager and Evans

Pondering in isolation upon the consequences of climate change, the imminence (or not) of the next Ice Age, or the timing of a future super-eruption or asteroid impact, can promote fleeting concern. Worse, consideration *en masse* of the future threats to our planet and our race is quite capable of contributing to bouts of severe depression. Let me summarize our current position. We are now well into a cycle of warming that is certain to lead to dramatic geophysical, social, and economic changes during the next hundred years, which will impinge—almost entirely detrimentally—on everyone. At the same time our planet is teetering on the edge of the next Ice Age, waiting in the wings should we ever come to our senses and tackle the remorseless rise in global temperatures. Some asteroids sufficiently large to wipe out a quarter of the human race continue to hurtle across the Earth's orbit undetected, while who knows when we will spot the next great comet heading our way. The Indian Ocean and Japanese tsunamis have provided us with glimpses of how terribly Nature can impact upon an increasingly crowded world, and there are now so many of us that

110

the next mega-tsunami or volcanic super-eruption cannot fail to result in millions of deaths and the enormous disruption of our so-called advanced global society. So interconnected is our social and economic framework that just a single quake in Japan could lead to worldwide economic disaster.

There are other trends too that will ensure an end to the world as we know it, and within this century. The world's population is still rising and is currently estimated at about 7.1 billion. According to the UN, by mid-century, numbers will have climbed to somewhere between 8.3 and 10.9 billion, and by a hundred years later, could be anything between 3.2 billion and a staggering 24.8 billion, dependent upon all sorts of different factors. Increasingly, as rampant climate change begins to bite and resources become ever more depleted, so our overcrowded world will, quite literally, become a battleground for fertile land, water, energy, and minerals.

Not only will the future of our descendants be hotter, more hostile, and increasingly impoverished of natural resources: it will also be dull and barren. In one of the greatest mass extinctions ever, our activities are currently wiping out between 3,000 and 30,000 species a year, from an estimated total of just 10 million. Up to one-third of flowering plants could be at risk, while between 25 and 50 per cent of all animal species could disappear before the chimes ring in the new century. As more and more species are obliterated their places will quickly be taken by the pests, weeds, and diseases that live cheek by jowl with the human race. Instead of a world of gorillas, pandas, birds of paradise, and corals, our descendants will have to make do with rats, cockroaches, thistles, and nettles. Furthermore, biodiversity is such a fragile thing, tenderly and incrementally reared by evolution, that it may take five million years or more for it to restore itself. In the meantime, we will have committed an estimated 500 trillion of our descendants to life on a dull and—in terms of variety—largely lifeless planet. Just as importantly, we are actually playing God with evolution itself and the entire future prospects of life on Earth. By wiping out many of the species that exist today, we are destroying much of evolution's raw material and severely limiting the planet's ability to generate the species of the future. For millions of species throughout geological time the end of the world has already come, and our activities are ensuring that the same fate will shortly face many of the life forms with which we currently share our planet.

The picture I am painting of the future, then, just seems to get worse and worse: a world of economic, political, and social upheaval, struggling—probably too late—to right the environmental wrongs wrought by their ancestors—us. Somehow, against this background, Brandon Carter's 'doom soon' scenario does not sound so far-fetched. Others too have come to the conclusion that things must come to a dramatic head soon. Anders Johansen and Didier Sornette of the University of California, for example, just over a decade ago predicted—on the basis of trends in population and economic and financial indices—that some sort of abrupt switch to a new 'regime' will occur around mid-century, the nature of which remains to be seen, but which is unlikely to be pleasant. Similarly, former UK Government Chief Scientist Sir John Beddington, and others, have warned of a 'perfect storm' developing around 2030, as a consequence of overwhelming demand and competition for food, water, and energy.

Probably the only piece of good news that can be taken away from my brief look at the end of the world as we know it, is that although this is going to happen—and soon—the survival of our race seems to be assured, for now at least. Leaving aside the possibility of a rare comet or asteroid impact on a scale of the dinosaur-killer, it is highly unlikely that anything else is going to wipe out all 7.1 billion of us in the foreseeable future. Even the replacement of the world with which we have become so familiar with one of sweltering heat might not seem as scary for those of our descendants likely to be in the thick of things. After all, we are a remarkably adaptable species, and can change to match new circumstances with some aplomb. Familiar 'worlds' have certainly ended many times before, as no doubt a centenarian born and raised while Queen Victoria sat on the throne of the United Kingdom, and who lived to see humans set foot on the Moon, would testify. The danger is that the world of our children and those that follow will be a world of struggle and strife with little prospect of, and perhaps little enthusiasm for, progress—as the Victorians viewed it. Indeed, it would not be entirely surprising if, at some future time when the great coastal cities are sinking beneath the waves, the general consensus held that there had been quite enough progress thank you—at least for a while.

Although I have tried, in these pages, to extrapolate current trends and ideas to tease out and examine somewhat depressing scenarios for the future of our planet and our race, I am sure that, to some extent at least, you would be justified in accusing me of a failure of the imagination. After all, I have rarely looked ahead beyond a few tens of thousands of years, and yet our Sun is still likely to be bathing our planet in its life-giving warmth for billions more. Who knows, over such an incomprehensible length of time, what Homo sapiens and the species that evolve from us will do and become. Our species and those that follow may be knocked back time and time again in the short term, but provided we learn to nurture our environment rather than exploit it, both here on Earth—before the Sun eventually swallows it up—and later, perhaps, in the solar system and the galaxy and beyond, then we have the time to do and be almost anything. Alternatively, maybe we can forget the science fiction and a human panspermy across the galaxy—à la *Star Trek*; the true future being Earth-bound, far less optimistic, and short. Could it be, as legendary animator and story-teller Oliver Postgate—of Ivor the Engine and Bagpuss fame—puts it in his wonderful autobiography, *Seeing Things*: that 'the real future is what is slowly coming up behind us, but we can't see it clearly because someone has stuck smiley faces over the wing mirrors'.

Further reading

The wide range of themes addressed in this small volume ensures that the breadth of further reading is huge. The list below represents, therefore, a tiny fraction of the thousands of books worth dipping into in order to find out more about specific areas that I have touched upon, and should be thought of more in terms of a personal choice than providing a balanced perspective.

Alvarez, Walter. *T. Rex and the Crater of Doom*. Princeton. 2008.

Berners-Lee, Mike, and Clark, Duncan. *2013 The Burning Question*. Profile Books. 2013.

Bryant, Edward. *Tsunami: The Underrated Hazard*. Cambridge University Press. 2001.

Emmott, Stephen. *10 Billion*. Penguin. 2013.

Fagan, Brian. *The Little Ice Age: How Climate Made History 1300–1850*. Basic Books. 2001.

Fagan, Brian. *The Complete Ice Age: How Climate Change Shaped the World*. Thames & Hudson. 2009.

Fortey, R. *The Earth: An Intimate History*. Harper. 2005.

Hamilton, Clive. *Earth Masters: The Dawn of the Age of Climate Engineering*. Yale University Press. 2013.

Lovelock, James. *The Vanishing Face of Gaia: A Final Warning*. Penguin. 2010.

McGuire, Bill. *Waking the Giant: How a Changing Climate Triggers Earthquakes, Tsunamis and Volcanoes*. Oxford University Press. 2012.

Musson, Roger. *The Million Death Quake: The Science of Predicting the Earth's Deadliest Natural Disaster*. Palgrave Macmillan. 2012.

Nield, Ted. *Incoming! Or Why We Should Stop Worrying and Learn to Love the Meteorite*. Granta. 2011.

Oppenheimer, Clive. *Eruptions That Shook the World*. Cambridge University Press. 2011.

Pearce, Fred. *The Climate Files: The Battle for the Truth About Global Warming*. Guardian Books. 2010.

Redfern, Martin. *The Earth: A Very Short Introduction*. Oxford University Press. 2003.

Walker, Gabrielle. *Snowball Earth: The Story of a Maverick Scientist and his Theory of the Global Catastrophe that Spawned Life as we Know it*. Bloomsbury. 2003.

Zalasiewicz, Jan, and Williams, Mark. *The Goldilocks Planet: The 4 Billion Year Story of Earth's Climate*. Oxford University Press. 2012.

Index

Global Catastrophes

Global Catastrophes

SOCIAL MEDIA
Very Short Introduction

Join our community
www.oup.com/vsi

- Join us online at the official Very Short Introductions **Facebook** page.
- Access the thoughts and musings of our authors with our online **blog**.
- Sign up for our monthly **e-newsletter** to receive information on all new titles publishing that month.
- Browse the full range of Very Short Introductions online.
- Read **extracts** from the Introductions for free.
- Visit our library of **Reading Guides**. These guides, written by our expert authors will help you to question again, why you think what you think.
- If you are a teacher or lecturer you can order inspection copies quickly and simply via our website.

ONLINE CATALOGUE
A Very Short Introduction

Our online catalogue is designed to make it easy to find your ideal Very Short Introduction. View the entire collection by subject area, watch author videos, read sample chapters, and download reading guides.

http://fds.oup.com/www.oup.co.uk/general/vsi/index.html

GLOBAL WARMING
A Very Short Introduction
Mark Maslin

Global warming is arguably the most critical and controversial issue facing the world in the twenty-first century. This *Very Short Introduction* provides a concise and accessible explanation of the key topics in the debate: looking at the predicted impact of climate change, exploring the political controversies of recent years, and explaining the proposed solutions. Fully updated for 2008, Mark Maslin's compelling account brings the reader right up to date, describing recent developments from US policy to the UK Climate Change Bill, and where we now stand with the Kyoto Protocol. He also includes a chapter on local solutions, reflecting the now widely held view that, to mitigate any impending disaster, governments as well as individuals must to act together.

363   McGuire, B.
.34   Global catastrophes.
McGu   Aurora P.L. APR15
2014   33164100083018